Henry Jacob Winser

The Great Northwest

A guide-book and itinerary for the use of tourists and travellers over the lines of the Northern Pacific Railroad, the Oregon Railway and Navigation Company

Henry Jacob Winser

The Great Northwest
A guide-book and itinerary for the use of tourists and travellers over the lines of the Northern Pacific Railroad, the Oregon Railway and Navigation Company

ISBN/EAN: 9783337190064

Printed in Europe, USA, Canada, Australia, Japan

Cover: Foto ©Andreas Hilbeck / pixelio.de

More available books at **www.hansebooks.com**

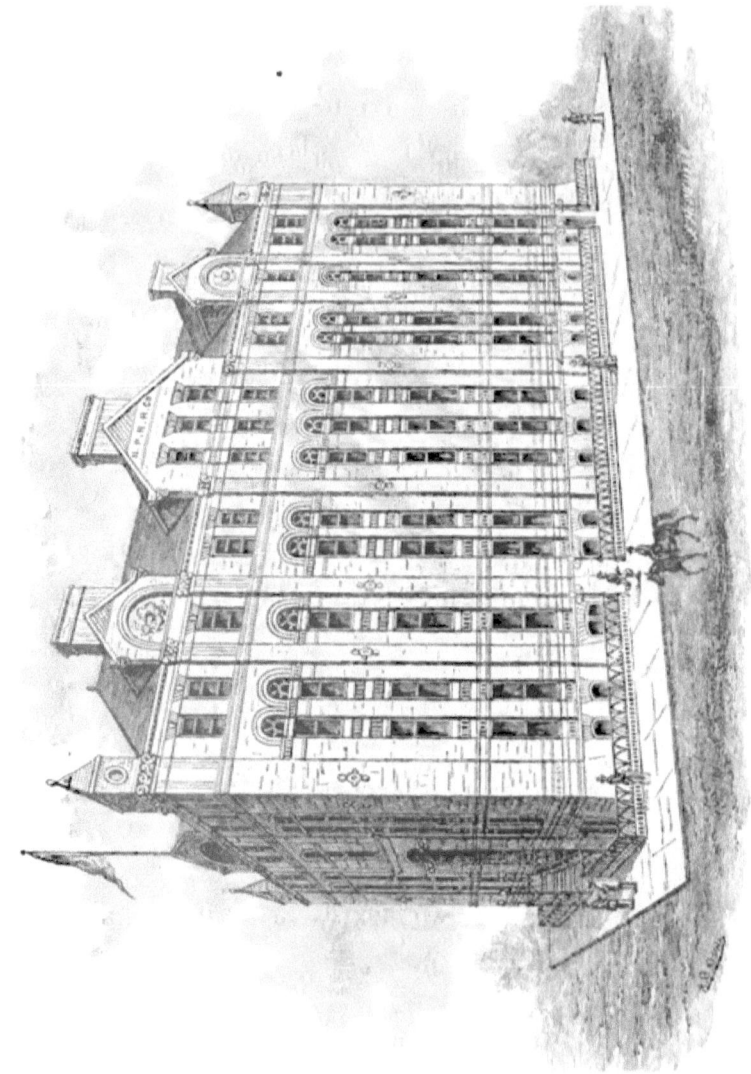

Headquarters and Offices, Northern Pacific R. R. Co., St. Paul, Minn.

THE GREAT NORTHWEST

A GUIDE-BOOK AND ITINERARY

FOR THE USE OF

TOURISTS AND TRAVELLERS

OVER THE LINES OF THE

NORTHERN PACIFIC RAILROAD

THE

OREGON RAILWAY AND NAVIGATION COMPANY

AND THE

OREGON AND CALIFORNIA RAILROAD

CONTAINING DESCRIPTIONS OF STATES, TERRITORIES, CITIES, TOWNS, AND PLACES ALONG THE ROUTES OF THESE ALLIED SYSTEMS OF TRANSPORTATION, AND EMBRACING FACTS RELATING TO THE HISTORY, RESOURCES, POPULATION, PRODUCTS, AND NATURAL FEATURES OF THE GREAT NORTHWEST

WITH MAP AND MANY ILLUSTRATIONS

BY

HENRY J. WINSER

Author of "The Yellowstone National Park ; a Manual for Tourists," etc.

NEW YORK
G. P. PUTNAM'S SONS
27 & 29 WEST 23D STREET
1883

*Press of
G. P. Putnam's Sons
New York*

CONTENTS.

	PAGES.
INTRODUCTORY	7–12
MINNESOTA	13–16
ST. PAUL	17–25
MINNEAPOLIS	26–34

ST. PAUL DIVISION:

St. Paul to Little Falls	35–38
Little Falls and Dakota Branch	39–43
Belle Prairie to Brainerd	43–48

WISCONSIN DIVISION:

Superior to Thompson	49–52

MINNESOTA DIVISION:

Duluth to Brainerd	53–61
Gull River to Wadena	61–65
Northern Pacific, Fergus and Black Hills Branch	66–72
Bluffton to Moorhead	72–83

NORTHERN DAKOTA	84–92
Fargo and South-western Branch Railroad	92–93

DAKOTA DIVISION:

Fargo to Jamestown	94–104
Jamestown and Northern Branch	105–106
Eldridge to Bismarck	106–112
The Great Bridge over the Missouri River	112–117
The Valley of the Upper Missouri	117–120

MISSOURI DIVISION:

Mandan to Scoria	121–130
The Bad Lands to the Boundary Line	130–136

MONTANA	137–141
THE YELLOWSTONE VALLEY	141–143

YELLOWSTONE DIVISION:

Glendive to Miles City	144–147
Explorations of the Yellowstone	147–148
A Fight with Indians at Tongue River, etc	148–159
The Massacre of Custer's Command	159–160
The Brilliant Work of General Miles	160–162
Fort Keogh to Pompey's Pillar	163–166
The Crow Indian Reservation	166–168

Contents.

MONTANA DIVISION:
- Billings..169–170
- Montana Stock and Sheep Raising.......................170–173
- Park City to Livingston.................................174–180
- Bozeman to Helena......................................180–184

ROCKY MOUNTAIN DIVISION:
- Helena to Butte, *via* Deer Lodge.....................185–193
- Garrison to Missoula..................................193–197
- The Flathead Indian Reservation........................197–200
- Paradise Valley and along Clark's Fork River..........200–204

PEND D'OREILLE DIVISION:
- Heron and Cabinet Landing................................205
- Idaho..206–207
- Lake Pend d'Oreille to Rathdrum.......................208–210
- Washington and Oregon.................................211–215
- Spokane Falls to Palouse Junction.....................216–219
- Palouse Junction to Wallula Junction..................219–221

OREGON RAILWAY AND NAVIGATION COMPANY:
- Wallula Junction to Bolles Junction...................222–224
- Bolles Junction to Dayton................................224
- Riparia to Lewiston...................................224–225

ALONG THE COLUMBIA RIVER:
- Wallula Junction to Umatilla..........................226–227
- Baker City Branch Line................................227–228
- The Main Line to The Dalles...........................228–235
- Dalles City to Portland...............................235–241

BY RIVER TO PORTLAND................................242–243

PORTLAND, OREGON....................................243–244

OREGON AND CALIFORNIA RAILROAD:
- Portland to Southern Oregon...........................245–253
- Portland to Corvallis (West Side Division)............254–256

THE OREGONIAN RAILWAY COMPANY......................257–258

A TRIP ON THE WILLAMETTE RIVER.....................259–260

NORTHERN PACIFIC RAILROAD—PACIFIC DIVISION:
- Portland to Tacoma....................................261–264
- Cascade Branch—Tacoma to Carbonado....................264–265
- The Seattle Branch—Seattle............................265–266

THE GLACIERS OF MOUNT TACOMA.......................267–271

THE LOWER COLUMBIA RIVER:
- Portland to Astoria...................................272–275

TRIPS ON THE SOUND AND ON THE OCEAN................275–276

ILLUSTRATIONS.

From Sketches by A. von Schilling, *and Photographs by* F. Jay Haynes, Davidson, *and others. Engraved by* E. Heinemann, A. Demarest, Matthews, Northrop & Co., *and* E. Clément.

Head-quarters and Offices, Northern Pacific Railroad Company, St. Paul, Minn.	Frontispiece
St. Paul, Minn.	17
Minneapolis, Minn.	26
Duluth, Minn.	53
Bonanza Wheat Farming—Plowing	84
Bonanza Wheat Farming—Seeding	87
Bonanza Wheat Farming—Harrowing	91
Bonanza Wheat Farming—Harvesting	96
The Northern Pacific Railroad Bridge over the Missouri River	112
Buttes in Pyramid Park	130
Pyramid Park Scenery	132
Eagle Cliff, near Glendive, Mont.	145
Buffalo Hunting in Eastern Montana	154
Current Ferry over the Yellowstone River	162
Big Horn River, Bridge and Tunnel	164
Indian Camp on the Line of the N. P. R. R.	167
Driving Cattle from the Range to the Railroad	173
Gates of the Mountains, near Livingston, Mont.	179
Three Forks of the Missouri River	180
Gates of the Rocky Mountains, near Helena, Mont.	186

Beaver Hill, Hell Gate Cañon, near Missoula, Mont.	195
Marent Gulch Bridge, Coriacan Defile	197
Alice Falls, Mont.	199
Cabinet Gorge, on the Clark's Fork River	205
Lake Pend d'Oreille, Idaho Territory.	208
Lake Cœur d'Alène, Idaho Territory	210
Spokane Falls, Washington Territory	215
Along the Cliffs of the Columbia.	236
Cascades of the Columbia River.	239
Multnomah Falls on the Columbia	240
Castle Rock on the Columbia	242
Cape Horn on the Columbia	244
Pillars of Hercules and Rooster Rock.	246
Portland, Oregon.	248
Iron Mountain, Cow Creek Cañon, Southern Oregon.	250
Tacoma, Washington Territory.	262
Seattle, Puget Sound, Washington Territory.	264
Distant View of Mount Tacoma.	266
Glaciers of Mount Tacoma.	268
Glaciers of Mount Tacoma.	270

INTRODUCTORY.

The region which is in process of development by the Northern Pacific Railroad, and the railroad systems with which it is in direct connection, embraces, in whole or in part, no less than seven of the largest States and Territories; *viz.*: Wisconsin, Minnesota, Dakota, Montana, Idaho, Washington and Oregon; or, at a rough estimate, one-sixth of the area of the United States.

The distance between the extreme eastern and western termini of the main line, on Lake Superior and Puget Sound, inclusive of 210 miles of railroad along the Columbia River which belong to the allied Oregon Railway and Navigation Company, is 2,168 miles.

The Northern Pacific Railroad is connected with the cities of St. Paul and Minneapolis by a lateral line, 136 miles in length. It has also various other branches, including one to the Yellowstone National Park, either finished or under construction, which represents a total of 700 miles of track. In addition to these branches, the trunk line has for its immediate tributaries the extensive systems of the Oregon Railway and Navigation Company, in Oregon and Washington Territory, and the Oregon and California Railroad, in Western Oregon.

This great system of allied railroads has opened to settlement during the past few years one of the fairest sections of the country—a region exceeded by no other part of the United States in its wealth of natural resources, nor surpassed in any of the conditions of climate or of soil which are best adapted to the well being of the human race.

The Great Northwest has already become famous for the prodigality of its cereal productions; the salubrity of its climate is an accepted fact; the extent and variety of its mineral deposits and the value of its grand forests are everywhere acknowledged, while the marked diversity and extraordinary attraction of its scenery are recognized as not the least prominent of its features.

Now that the Northern Pacific Railroad is finished, the inviting regions of the Great Northwest, hitherto remote, are made easy of access. The tide of travel will flow naturally with a strong current through this new and pleasant channel, and to pilot the wayfarer this Guide Book has been written.

The aim has been to furnish the tourist and traveler with precisely that information which would seem requisite through the successive stages of the journey. The book embraces facts with reference to the history, present population, productions, resources and natural features of the country traversed by the Northern Pacific Railroad and its branch lines, and by its allies, the Oregon Railway and Navigation Company, and the Oregon and California Railroad, with some account of the ocean and river routes of the Pacific Northwest.

The salient features of the states, territories, cities, towns, and all places of interest along the lines of these vast systems of railroad and water transportation are described, and such material of local character is interspersed among the pages which may serve to interest the traveler in the course of his journey. In collating the facts which are here given to the public, the author has spared no effort to secure the utmost freshness and accuracy. The growth of the Great Northwest—its cities and towns—in population and material prosperity is, however, so rapid, that the figures of to-day may seem far short of the truth a few months afterward.

In this connection thanks are due to the officials of the various railroad organizations, and also to many private individuals, for contributions of valuable data. From Government and other reports, the historical features of the work have been procured. By far the largest part of the book, however, is the record of personal investigation, supplemented, as far as relates to the Pacific Northwest, by information obtained from Mr. S. A. Clarke, of Portland, Oregon, whose residence of thirty-five years on the Pacific Coast has given him a vast store of knowledge on all matters pertaining to that region.

Outline of the Northern Pacific Railroad's History.—The charter and organization of the Northern Pacific Railroad Company date from 1864, but the project to build the railroad over substantially the same route now traversed by the company's main line is much older. Indeed, it is the oldest of all projects to open railway communication with the Pacific coast.

Introductory.

A railroad from the upper Mississippi to the mouth of the Columbia River was advocated as long ago as 1835, soon after the railway system was introduced in this country. About ten years later an enterprising New York merchant, named Asa Whitney, who had made a fortune in China, urged upon Congress, session after session, a plan for building a railroad from the head of Lake Michigan, or from Prairie du Chien, on the Mississippi River, to the mouth of the Columbia River, in Oregon. He asked a land grant of sixty miles in width along the whole line of his proposed route. Many State legislatures passed resolutions in favor of Whitney's project, and Congress gave it much serious consideration. At one time Whitney's bill was within one vote of passing the Senate.

After the Mexican war came the annexation of California, followed by the gold discoveries and the rapid growth of population in that State. Then the general opinion in Congress and the country naturally favored the building of the first transcontinental line of railroad on a route ending at the Bay of San Francisco. Accordingly, the Union and Central Pacific companies were chartered in 1862, with a grant of public lands and a large subsidy of Government bonds. Among the projectors of a line to California was Josiah Perham, of Maine, then living in Boston, who had a charter from the State of Maine for the People's Pacific Railroad Company, and who, in vain, attempted to get Congress to adopt his company, and give it the grants subsequently given to the Union and Central companies. Failing in this effort, Mr. Perham turned to the Northern road, which had been long and ably advocated as the best line to the Pacific coast by the eminent engineer, Edward F. Johnson, and by Gov. Stevens, of Washington Territory, who had been in command of the Government expedition that surveyed the northern line in 1853. Stevens' surveys had shown the Northern road was not only feasible, but was a better line in respect to grades and in regard to the character of the country traversed than any other.

In 1864 Congress passed a bill chartering the Northern Pacific Railroad Company, and naming as incorporators among others the men concerned with Perham in the old abortive People's Pacific Company. Under this charter the company was organized in Boston, with Mr. Perham as its President, and an attempt was made to raise money for the construction of the road by a popular subscription to shares of stock at $100 each. This attempt was an absolute failure, and after a year's futile effort Mr. Perham and his associates turned over the charter of the company to an organization of New England capitalists and railroad men, who proposed to make the road tributary to Boston. They elected J. Gregory Smith, of the

Vermont Central Railroad, President of the Northern Pacific Company. Smith and his associates tried in vain for several years to obtain legislation from Congress guaranteeing the interest on the company's stock. The original charter did not allow the issue of bonds. Attempts in this direction were abandoned in 1869, and amendments to the charter were procured allowing the company to mortgage its road and land grant. A contract was then made with the banking house of Jay Cooke & Co., of Philadelphia, to sell the company's bonds. Mr. Cooke had negotiated the great war loans of the Government, and was regarded as the most successful financier in the country. In the short period of about two years, his firm disposed of over thirty millions of dollars of Northern Pacific bonds, bearing interest at $7\frac{3}{10}$ per cent. With the money thus obtained, the work of construction was begun in the spring of 1870, and by the fall of 1873 the road had been completed from Duluth, at the head of Lake Superior, to Bismarck, on the Missouri River, and from Kalama, on the Columbia River, in Washington Territory, to Tacoma, on Puget Sound, the total number of miles completed being about 600.

The great financial panic of 1873 prostrated the house of Jay Cooke & Co., wholly stopped the sale of Northern Pacific bonds, and made it impossible to go on with the road. The company was insolvent, and after a time its directors threw it into bankruptcy, and with the cordial assent of its bondholders reorganized its affairs so as to free it from debt, by converting its outstanding bonds into preferred stock. When the effects of the panic and the succeeding hard times had begun to pass by, the managers of the Northern Pacific recommenced the work of building its long line across the Continent. The construction began with the Cascade branch, from Tacoma to the newly discovered coal fields at the base of the Cascade Mountains. Then a loan was negotiated for building the Missouri Division, from the Missouri to the Yellowstone rivers, and shortly afterwards another loan for the construction of the Pend d'Oreille Division from the mouth of the Columbia River to Lake Pend d'Oreille, in Idaho. In the meantime, several changes had occurred in the Presidency of the road. President Smith had been succeeded in 1874 by General Cass, and he by Charles B. Wright, of Philadelphia. Mr. Wright's resignation in 1879 was followed by the election of Frederick Billings, under whose management the work of construction was carried on until 1881. A general first mortgage loan was negotiated to provide the means for completing and equipping the entire line. The credit of the company had by this time become so good, that its bonds were readily sold above par by a syndicate of the leading bankers of New York city.

In 1881, Henry Villard, who had previously obtained control of all the transportation lines, both rail, sea and river, in Oregon and Washington, purchased for himself and friends a controlling interest in the stock of the Northern Pacific Company, and was elected its President. His purpose was to ally to the Continental Trunk Line, as feeders and extensions, the lines under his management on the Pacific Coast. To accomplish this and to secure an identity of interest, he organized the Oregon and Transcontinental Company, which holds a large portion of the stock of the Northern Pacific, the Oregon Railway and Navigation Company, and the Oregon and California Railway Company, and which builds branches for the Northern Pacific under an arrangement by which the latter company operates them and in time becomes the owner of their stock. Under the efficient management of President Villard, and Vice-President Thomas F. Oakes, the work on both ends of the Northern Pacific was prosecuted with great vigor during the years 1881, 1882 and 1883, until the ends of track advancing from both sides of the Continent met near the summit of the Rocky Mountains in the fall of 1883.

With its main line and branches, the Northern Pacific now operates over 3,000 miles of railroad, and traverses the only continuously habitable belt of country stretching across the American continent from the valley of the Mississippi to the Pacific Ocean. Its future must be a brilliant one. Its management is conservative and prudent, aiming to make its road a powerful agency in the settlement and prosperity of the whole vast region which it drains.*

Routes from the East to St. Paul.—The distances from New York, via Chicago to St. Paul, Minn., by the several Trunk lines, are as follows:

NEW YORK TO CHICAGO.

	MILES.
Via Pennsylvania Railroad	912
" Erie Railway	958
" New York Central Railroad	977
" Baltimore and Ohio Railroad	1,041

* For a complete history of the Northern Pacific Railroad, the reader is referred to a volume recently published by G. P. Putnam's Sons, New York, which describes the beginning and progress of the enterprise, its legislative, financial and administrative phases, the engineering and constructive work on the line, and the resources and chief characteristics of the extensive regions traversed by the road.

CHICAGO TO ST. PAUL.

Via Chicago, Milwaukee and St. Paul Railway............ 410
" Chicago and Northwestern Railway................... 410
" Chicago, Rock Island and Pacific Railway............ 529

Tourists from Philadelphia usually go by way of the Pennsylvania Central. From Baltimore and Washington there is choice of either the Northern Central, with Pennsylvania Central connections, or the Baltimore and Ohio Railroad. From Boston, the most direct route is that of the Boston and Albany Railroad to its connection with the New York Central.

Via the Lakes.—In summer, tourists may take a comfortable and agreeable method of reaching the Northern Pacific Railroad at Duluth, on Lake Superior, by way of the great lakes. The Lake Superior Transit Company runs lines of fine passenger steamers from Buffalo, Cleveland, Detroit and other ports, through Lake Erie, Lake Huron, and by way of the great Government Canal of Sault Ste. Marie, passing through the whole length of Lake Superior and touching many of its interesting ports.

Sleeping Car Expenses.—Pullman, or other palace cars, run on the through trains of all the aforementioned railroads. The cost of sleeping car accommodation is the same on all the routes, and the tariff is as follows:

	HALF SECTION.	SECTION.
From New York to Chicago.................	$5 00	$10 00
" Chicago to St. Paul.................	2 00	4 00
" St. Paul to Portland.................	8 00	16 00

The Northern Pacific Railroad is equipped with Pullman Palace Cars of the latest construction, the appointments of which combine the newest inventions for the perfect accommodation and absolute comfort of passengers.

Dining Cars.—Dining cars, built expressly to meet the needs of the long journey, are attached to all through trains of the Northern Pacific Railroad. In these cars sumptuous meals are served at the uniform rate of seventy-five cents. This adds greatly to the luxury of traveling, entirely obviating the discomfort which is too often experienced where dependence for food is placed upon wayside dining stations.

Transfer Coaches. In all the Western cities there are lines of transfer coaches ready on the arrival of the train to take the traveler and his baggage direct to any hotel, or transfer him across the city to any depot. The transfer agent passes through the cars before the arrival of the train, selling transfer checks and tickets. The service is trustworthy and convenient, and the charge is uniformly fifty cents.

CORRECTION—SEE P. 12.

Sleeping Car Expenses.—Pullman or other palace cars run on the through trains of all the aforementioned railroads. The cost of sleeping car accommodation is the same on all the routes, and the tariff is as follows:

	HALF SECTION.	SECTION.
From New York to Chicago	$5 00	$10 00
" Chicago to St. Paul	2 00	4 00
" St. Paul to Portland	15 00	30 00

Minnesota.

Minnesota is situated in a high northern latitude, elevated from 1,000 to 1,800 feet above the ocean, and has a peculiarly dry and salubrious climate. The most rugged portions of the State are about Lake Superior, along the Mississippi River and on the heights of land which divide the sources of the great river systems. The area of Minnesota embraces 83,530 square miles, or 53,459,840 acres, about two-thirds of which is undulating and well adapted to cultivation. Within the limits of the State three great river and lake systems have their sources, viz.: the Mississippi and its northern tributaries; the St. Louis River and its numerous branches, forming the head of those waters which find their way through the great lakes into the St. Lawrence; and lastly, the affluents of the Red River of the North, Rainy Lake and Lake of the Woods, which discharge their waters into Hudson's Bay. Wood-girt lakes, more than 7,000 in number, most of which are quite deep and full of fish, having shores generally firm and dry, bottoms sandy or pebbly, and waters clear, cool and pure, dimple in every direction the undulating lands. The average area of these lakes is about three hundred acres, but many are very large. For example, Red Lake is estimated to equal 340,000 acres; Mille Lacs, 130,000; Leech Lake, 114,000; Winnebagoshish, 56,000; Minnetonka, 16,000, and a number of others exceed 5,000 acres.

The Mississippi, rising in Lake Itasca, 826 feet above the mouth of Lake Minnesota, lends the State a shore line of one thousand miles. In its descent, the river is broken at intervals by numerous falls and rapids which afford valuable water powers. Among the more important of those which have been utilized are the Falls of St. Anthony, at Minneapolis, one of the largest water powers in the known world; Sauk Rapids, with 32,000 horse-power; Pike Rapids, 12,000; Prairie Rapids, 6,000; Olmstead's Bar, 9,000, and French Rapids, above Brainerd, 670. Nineteen of Minnesota's streams pour their waters into the Mississippi, each of which affords water power of more or less capacity.

The St. Louis River, rising in the northern part of St. Louis County, flows through a vast pine region, and after receiving the waters of many tributary streams descends to the level of Lake Superior. There is an immense water power afforded by the numerous rapids in this river. Its fall is estimated at over 500 feet during a course of fifteen miles, and the power is believed to be equal to that of the Mississippi at Minneapolis. A company has been recently organized with a capital of $1,000,000, having in view the development of the water power on this river.

Fergus Falls, on the Red River, and several other falls and rapids on the Knife, Cloquet, Moose, Kettle, St. Croix, and a score of other fine streams, exhibit the distribution of water power throughout the State.

Only a small fraction of this power has been developed or is now in use; but, considering its magnitude and diffusion, the capacity of the surrounding country for supplying the raw material, and the widespread field for the consumption of manufactured products, it is impossible to limit the industrial progress which this bounteous water power makes possible.

No Western State has made more progress in railroad construction than Minnesota, and none possesses greater facilities for travel and transportation. At the end of 1862 there were only ten miles of railroad in operation; but twenty years afterwards, at the close of 1882, a network of over 4,000 miles of railroad covered the State, bringing every town and village, except those in the great unorganized counties of the northern section, within twenty miles of a railroad station. As to navigable waters, there are not less than 2,796 miles of shore line within the limits of the State, or about one mile of coast line to every thirty square miles of surface.

The soil of Minnesota is very fertile, and the increase of agricultural production has kept pace with the development of railways and other means of transportation. The manufacturing interests of the State are already very large — flour and lumber being the leading commodities, although there are a great variety of other important industries. The products of the mills alone were estimated in round numbers at $100,000,000 in value in 1882. The population, at present about 1,000,000, is largely increased every year, and the steady advance in the taxable valuations of property shows that the commonwealth is rapidly growing in material prosperity.

Although Minnesota is generally classed as a prairie State, in reality its surface is about one-third wooded. The timber lands of Minnesota extend over a large part of the northern and eastern sections of the State. The hard wood belt alone covers an area of 5,000 square miles, and consists of white and black oak, maple, hickory, elm, with varieties of soft woods, such as spruce, tamarack and cottonwood.

The pine lands stretch over an immense territory, clothing the head waters of the great river and lake systems which have their sources in Minnesota, and forming a continuation of the great pine belt which extends across northern Wisconsin. This great tract of timber is traversed by the waters of the St. Croix, Mississippi, St. Louis, and their myriad of tributaries, thus furnishing convenient channels for floating the logs which are cut during the winter, at high water in the spring, to points where saw mills convert them into lumber.

From the head of Lake Superior west 150 miles, and east from Superior City along the Wisconsin Division of the Northern Pacific Railroad, now in course of construction to Montreal River, the boundary line between Wisconsin and Michigan, the railroad traverses a country of great wealth, consisting of densely wooded and magnificent pine forests interspersed with occasional tracts of hard wood timber. The pine lands in Minnesota extend north of the line of the Northern Pacific Railroad a distance of over one hundred miles, covering an area of more than twenty thousand square miles, and afford a veritable gold mine to the lumberman and manufacturer.

An idea of the importance of the lumber industry may be obtained from the fact that the product of the mills at Minneapolis, in 1882, amounted to 314,362,166 feet of lumber, 138,546,000 shingles and 61,330,380 laths, while that of Duluth during the same year was 84,218,153 feet of lumber, 21,363,000 shingles and 10,295,000 laths, being a very large increase respectively over that of the previous year. The total product of the lumber mills situated on the lines of the Northern Pacific Railroad in Minnesota, during the year 1882, amounted to 596,855,000 feet of lumber, 232,649,000 shingles and 105,950,380 laths.

In the description of Minneapolis will be found an account of the flouring mill industry, which is centred at that city.

The climate of Minnesota possesses those characteristics which are peculiar to the northern belt of the temperate zone at a considerable distance from the sea-board. The range of the thermometer is great at all seasons, frequently exceeding 50° during the winter and spring months, and showing variations of 40° in the summer season. For six years, the mean winter temperature, as given by the United States Signal Service at St. Paul, was 18° 45' in winter, 45° 50' in spring, 70° 49' in summer, and 44° 14' in autumn, and this included two remarkably cold seasons. The heat in summer is almost tropical, forcing vegetation with great rapidity and luxuriance. The thermometer in winter often drops under zero, sometimes registering 30° below, but the stillness and dryness of the air make the cold far from disagreeable. An ordinary still day in Minnesota, with the thermometer ranging from zero to 10° or 12° below, is really enjoyable, and mechanics are able to work out of doors at this temperature without inconvenience. Spring does not linger in the lap of winter, but bursts forth on the approach of May; and the Indian summer, late in November, is a season of almost magical beauty and softness. The climate, indeed, is considered one of the most healthy in the world. Persons afflicted with pulmonary diseases are sent to Minnesota to recover their strength and vigor, and thousands of consumptive patients bless the dry and balmy qualities in the atmosphere, which are potent enough to rescue such sufferers from untimely death.

J. M. Young
Thornhill
Galt.

James Young.

Boucher de LaBruère
　　　　　S. Hyacinthe

Joseph Lassé
　　　Montréal

John M. Martin
　　　Toronto

St. Paul, Minnesota.

St. Paul.

This city, the capital of Minnesota, is picturesquely situated at the head of steamboat navigation on the Mississippi River, over two thousand miles from its mouth. Thirty-four years ago, when Congress gave the State its territorial organization, fixing the seat of government at St. Paul, the name of the place appeared on no map, and geographers only knew that it was a very small settlement, somewhere near the Falls of St. Anthony, in that indefinite region called the Far West. To-day St. Paul has nearly 80,000 inhabitants, and its wonderful growth and prosperity fairly class it among the remarkable cities of the great Northwest. Many of its people, now scarcely beyond middle age, remember well its appearance on the advent of the white man's civilization. It was at that time a favorite resort of the Indians, to whom it was known as Im-mi-gas-ka, or White Rock, on account of the towering bluffs of sandstone which mark the course of the river. A succession of undulating and beautiful hills, clothed with forest, overlooked the Father of Waters, whose banks were bordered by graceful elms. The valleys between the heights were little more than deep ravines, through which numbers of rivulets flowed down to the great river. The site of St. Paul has been recently described by one of the pioneers, who reached the place in 1853, "as showing here and there a log shanty inhabited by the white men who had ventured to the head waters of the Mississippi River in search of adventure or gain, but by far the bulk of the population was represented by the untutored savage, whose tepees and wigwams occupied the hills and valleys that now constitute the city. Since that time a generation has not passed away, and behold what man hath wrought! The writer, who still claims to be a young man, *crawled* into the embryo city in 1853, being landed at the lower levee, near what is now the foot of Jackson Street. The only means then provided for getting into the city proper were a number of steps cut into the bluff, the top of which at that time would enable a person to step into the third story window of the present St. Paul Fire and Marine Insurance Building, on the corner of Third and Jackson

Streets. A few feet distant from the top of the bluff was a log hotel, a story and a half high, which was then known as the Merchants' Hotel, the building now on its site still retaining that name. A little further up, at the present crossing of Fourth and Jackson Streets, was a rude bridge which spanned a ravine some twenty-five or thirty feet deep, through which during certain seasons of the year a rushing torrent of water found its way to an outlet in the Mississippi River near Dayton's Bluff. This stream had its rise in and was the outlet of what was then a lake or marsh at the foot of St. Anthony Hill, upon the bosom of which it was not an unusual sight to see all kinds of water fowl, and there are persons still living in St. Paul who have shot wild geese and ducks there. At that time, during the season of high water, steamboats of the largest size would enter the outlet of this stream near Dayton's Bluff, and land within a few hundred feet of the Merchants' Hotel, sailing over the space that is now occupied by solid stone and brick blocks, where almost the entire wholesale business of St. Paul is at present conducted. The rushing torrent has given place to paved streets, and the lake, with fifteen to twenty feet depth of water, is now covered with four and five story business blocks of buildings. It is within the memory of the comparatively new comers to St. Paul when a traveling circus company spread its canvas upon Baptist Hill, fully twenty feet above the highest chimney of the substantial block on the corner of Fourth and Sibley Streets. Third Street was but a sort of straggling highway. The locality now occupied by the fine block running from Third to Fourth Street on Minnesota Street was the burial place of the dead. The old capitol building, which was destroyed by fire in 1881, was then in course of erection, and to reach it the pedestrian had to wade through mud nearly knee-deep, without a sign of a sidewalk, or any attempt at street grading. St. Paul proper, as it is now designated upon the maps, was about all there was of any attempts made toward laying out even a village."

St. Paul is built upon a succession of four distinct terraces, which rise in gradation from the river. The first is the low bottom which forms the levee. This was formerly subject to overflow; but it has been raised above high-water mark, and is now a very valuable property, occupied by warehouses, railroad tracks, the Union Depot and business offices. On the second and third terraces the principal part of the city is established. The second terrace, which is about ninety feet above the level of the river, is also devoted to business, and is thickly studded with fine blocks of buildings. Some of these are so commandingly situated on the high bluffs which overhang the Mississippi as to be visible a long distance up and down the stream, giving the city an imposing architectural appearance as it is approached by rail or river. The third terrace, very little higher than the second, widens out into a broad plateau, upon which stands much of the residence portion of the city. These upper terraces are on a foundation of blue limestone rock, from twelve to twenty feet in thickness, forming an excellent building material. Beneath this stratum is a bed of friable white quartzose sandstone of unknown depth, which is easily tunneled, and through which all the sewers have been excavated. The fourth, or highest terrace, is a semicircular range of hills, enclosing the main portion of St. Paul as in an amphitheatre. The picturesque sweep of these heights, conforming to the curve of the river, with their growth of native forests, and the stately residences which are scattered over their slopes, is a characteristic charm of St. Paul. Fine avenues have been laid out over many of the hills, leading away into the prairie lands beyond, or to some of the beautiful lakes in the neighborhood, and the residence part of the city is rapidly extending in every direction.

It cannot be denied that the site of St. Paul was a costly one upon which to build a city. Its hills had to be leveled, its valleys filled up and its crookedness made straight, at great expense of labor and money. But the street improvements are carried on regardless of the necessary expense, and the result already is eminently satisfactory. The capital of Minnesota is likely always to be noted as much for its beauty and salubrity as for its enterprise and commercial prosperity. The inconveniences and discomforts of frontier life have long since disappeared. The streets are paved and sewered, and lighted with gas and electricity. Pure water, in ample supply, is brought to every house through miles of pipe from distant lakes; street cars traverse the city in all directions; frequent local trains give easy access to the charming suburbs; the rough log chapel of 1849 has been superseded by forty-five places of worship, many of which are beautiful in architecture; the primitive log school-house has given place to fourteen massive public school buildings of brick and stone, in which are educated more than 7,000 pupils, at a yearly cost to the city of $120,000, with a valuable adjunct in parochial and private schools that furnish instruction to 2,500 scholars besides; there are three hospitals, two orphan asylums, a dozen banks, many imposing public and private buildings, several charitable and social institutions, excellent vocal and instrumental musical organizations, a boating club, lodges of secret and mutual benefit associations, military organizations, an opera house in which the brightest stars in the theatrical firmament from time to time appear to refined and fashionable audiences, public libraries and ab'y conducted newspapers, the *Pioneer-Press* being the leading journal of the Northwest.

Among the noteworthy buildings are the City Market, the State Capitol, the Post-office and Custom House, all on Wabasha Street; several fine business blocks on Third, Fourth, Jackson and other streets, and many of the churches. The principal hotels are the Metropolitan, the Merchants' and the Windsor, but there are several others, scarcely inferior in size, and not less well conducted.

The wholesale trade of St. Paul at the date of the latest statistics (1881) amounted to $51,000,000 for that year, and has since greatly increased. The manufacturing interest at the same date comprised 792 distinct establishments, which employed 9,000 workmen and produced $16,000,000 worth of goods during the year. These manufacturing industries are of quite recent growth, and bid fair to develop a great addition to the population and wealth of the city. St. Paul being a port of entry, many of her merchants import direct from foreign countries without the intervention of seaboard cities, and thus are enabled to supply their customers with certain kinds of foreign goods at cheaper rates than if bulk had to be broken at the port of first arrival.

Few cities possess so great facilities for travel and transportation as St. Paul. Here is the centre of the railroad systems of the great Northwest. Beside the Northern Pacific Railroad, there are not less than five trunk lines terminating in the city, bringing it in direct communication with the remotest parts of the United States and Canada, and penetrating the surrounding country in every direction with their branches. One hundred and fifty trains arrive at the Union Depot each day, and the same number depart, every one of which is freighted with the traffic of an empire that pours its tribute into the lap of the city. St. Paul lies midway between the Atlantic and Pacific Oceans, and its growth is stimulated from the east and from the west. The head-quarters of many of the railroad organizations are situated here, the large and convenient building of the Northern Pacific Railroad Company on Fourth Street being strikingly notable in point of architectural beauty. The St. Paul, Minneapolis and Manitoba, the Chicago, St. Paul, Minneapolis and Omaha, the Minneapolis and St. Louis, the Chicago and Northwestern, and the Chicago, Milwaukee and St. Paul Railroads have also erected elegant and commodious buildings in which to transact their business. Here, also, are railroad shops and transfers, acres of tracks, and elaborate depots to accommodate the enormous and rapidly growing traffic of the city. The activity in railroad extension on the part of all the trunk lines is constant, because as soon as new regions are made accessible the tide of immigration is sure to overflow them, and thus is building up a robust agricultural, commercial and manufacturing life, with St. Paul as a pivotal point, with a rapidity for which history affords no parallel.

St. Paul also does a large shipping business by way of the Mississippi River. During at least six months of the year fine passenger and freight steamers ply between the city and St. Louis, affording travelers an enjoyable trip by water, and giving cheap rates for merchandise. It is hoped that when the Government has finished its improvements in the channel of the river, wheat and other produce may be transported by steamer direct to New Orleans, and thence be shipped to Liverpool at lower rates than by rail.

The social life of St. Paul has lost every trace of provincialism, and in point of culture and refinement will compare favorably with that of far larger cities. On this point said a recent writer: " In the churches, the schools, the literary and social organizations, and also in politics, the most advanced and meritorious ideas and principles form the foundation, and the laws, political and social, are administered with rare discretion, judgment and grace. The business community consists of young, active men, who have mostly been schooled in mercantile pursuits in the Eastern States, but who find here opportunities for expansion and development not offered in their old homes. With a vaster field in which to operate, and an active, energetic class of men with whom to compete in traffic, their minds assume a broader scope, and their faculties become sharpened and more acute. These remarks apply not only to St. Paul, but to the entire new Northwest, the same spirit and attributes being found even in the smallest places. Young men of good principles and industrious habits always succeed here—success is in the climate, and wealth is in the soil."

There are many beautiful drives in the city and its suburbs, and a large number of resorts in the neighborhood, which may be reached by river and rail. The drive to *Lake Como*, four miles distant, is over a fine, hard gravel road, and the jaunt thither on a cool summer evening is delightful. In Dayton's Bluff, near the river, on the east side of the city, is the natural curiosity known as *Carver's Cave*, named after Jonathan Carver, of Connecticut, who, in 1763, under a commission from the King of England, led an exploring expedition into this region, and made a treaty with the Indians, by which the title to an immense tract of land was ceded to him. Carver described the country as being beautiful, the soil fertile, and the climate agreeable, and proposed founding a colony, but his designs were frustrated by the breaking out of the war of American Independence.

White Bear Lake, twelve miles distant, on the St. Paul and Duluth Railroad, is about nine miles in circumference. Its picturesque shores are lined with summer hotels and beautiful villas, and a large wooded island, recently connected with the main land by a causeway and bridge, has been laid out by the wealthy residents of St. Paul into plats for summer residences. The lake affords excellent fishing, boating and bathing. *Bald Eagle Lake*, a mile beyond, noted for its scenery and good opportunities for fishing, is quite popular as a resort for picnic parties. *Lake Elmo*, on the Chicago, St. Paul, Minneapolis and Omaha Railroad, twelve miles eastward of St. Paul, is also a much frequented summer resort, offering great attractions for boating, bathing and fishing. The hotel, accommodating 200 guests, is usually tested to its full capacity during the season. The water of these lakes is very bright and pure, and the shores are forested with oak, maple, elm, hickory and other deciduous trees.

Fort Snelling.—This military post, the head-quarters of the Department of Dakota, was established in 1819, with the view of protecting the few settlers who, at so early a date, were brave enough to penetrate the great wilderness west of the Mississippi. The fort is massively built on the northern bank of the Minnesota River, just at its junction with the Father of Waters. The situation of the fort is strikingly picturesque, its white walls reared upon the brink of a jutting bluff with an almost vertical face, its base being washed by the flood one hundred feet below. Fort Snelling was finished in 1822. Its form was circular, and its high walls were broken at intervals by embrasures for cannon to sweep the approaches. It has since undergone some alterations, but the original structure still remains. This fort has had an eventful history, having witnessed many scenes of savage warfare. The post is now used as a rendezvous for troops on the way to the frontier, only a few companies being permanently in garrison. Brig.-Gen. Alfred H. Terry is in command. His residence is in the centre of a long line of detached villas, recently erected by the Government for the convenience of the officers of the department. Fort Snelling is about half way between St. Paul and Minneapolis, being connected with the main road by a long iron bridge which airily spans the Mississippi.

The Falls of Minnehaha.—This beautiful water-fall, made immortal by Longfellow in his poem, "Hiawatha," is to be seen on the road toward Minneapolis, two miles beyond Fort Snelling. It is formed by an abrupt break in the bed of Little Minnehaha Creek, one of the outlets of Lake Minnetonka. This stream babbles along through miles of verdant meadows in the most quiet and commonplace way, to make an unexpected leap at last into a deep gorge, and find itself famous and beautiful. In a recent issue of *Harper's Magazine* the Falls of Minnehaha are aptly characterized by Ernest Ingersoll in this wise:

"The outlet of Lake Minnetonka is a sparkling little brook that encircles the city, steals through the wheat fields, races under a dark culvert where the phœbe birds breed, and then, with most gleeful abandon, leaps off a precipice sixty feet straight down into a maple-shadowed, brier-choked cañon, and prattles on as though nothing had happened but a bit of childish gymnastics.

"It is very charming, this rough and rock-hemmed little gorge through the woods and fern-brakes, and this fraudulent little beauty of a cascade; and it laughs without a prick of conscience, laughs in the most feminine and silvery tones from a rainbow-tinted and smiling face, when you remind it that it is a bewitching little thief of credit—for the true Minnehaha is over on the brimming river, a slave to the mills. But, right or wrong, little stream, thou art a princess among all the cascades of the world. Thy beauty grows upon us, and lingers in our minds like that of a lovely child, whether we wade into the brown water at thy feet, scaring the happy fishes clustered there, and gaze upward at the snowy festoons that with a soft, hissing murmur of delight chase each other down the swift slope; or creep to thy grassy margin above, and try to count the wavelets crowding to glide so glibly over the round, transparent brink; or walk behind thy veil, and view the green valley as thou seest it, through the silvery and iridescent haze of thy mist drapery. Thou hast no need of a poet's pen to sing thy praise; but had not the poet helped thy fraud, enchanting Minnehaha, not half this daily crowd would come to see thee, and to drink beer on thy banks, and murmur maudlin nonsense about Hiawatha and his mystical maiden. Nevertheless, thou art the loveliest of cascades, and an enchantress whose sins can be forgiven because of thy beauty."

In his entertaining book, "Life on the Mississippi," Mr. Samuel L. Clemens, known so widely as the clever writer, "Mark Twain," describes St. Paul in so piquant a way that we quote his words, although the city has long since outgrown the author's statistics:

"St. Paul is a town. It is put together in solid blocks of honest brick and stone, and has the air of intending to stay. Its post-office was established thirty-six years ago; and by and by, when the postmaster received a letter, he carried it to Washington, horseback, to inquire what was to be done with it. Such is the legend. Two frame houses were built that year, and several persons were added to the population. A recent number of the leading St. Paul paper, the *Pioneer Press*, gives some statistics which furnish a vivid contrast to that old state of things, to wit: Population, autumn of the present year (1882), 71,000; number of letters handled, first half of the year, 1,200,587; number of houses built during three-quarters of the year, 989; their cost, $3,186,000. The increase of letters over the corresponding six months of last year was fifty per cent. Last year the new buildings added to the city cost above $4,500,000. St. Paul's strength lies in her commerce—I mean, his commerce. He is a manufacturing city, of course—all the cities of that region are—but he is peculiarly strong in the matter of commerce. Last year his jobbing trade amounted to upwards of $52,000,000. He has a custom-house, and is building a costly capitol to replace the one recently burned—for he is the capital of the State. He has churches without end; and not the cheap, poor kind, but the kind that the rich Protestant puts up, the kind that the poor Irish 'hired girl' delights to erect. What a passion for building majestic churches the Irish hired girl has! It is a fine thing for our architecture; but too often we enjoy her stately fanes without giving her a grateful thought. In fact, instead of reflecting that 'every brick and stone in this beautiful edifice represents an ache or a pain, and a handful of sweat, and hours of heavy fatigue, contributed by the back and forehead and bones of poverty,' it is our habit to forget these things entirely, and merely glorify the mighty temple itself, without vouchsafing one praiseful thought to its humble builder, whose rich heart and withered purse it symbolizes. This is a land of libraries and schools. St. Paul has three public libraries, and they contain in the aggregate some 40,000 books. He has 116 school-houses, and pays out more than $70,000 a year in teachers' salaries. There is an unusually fine railway station; so large is it, in fact, that it seemed somewhat overdone, in the matter of size, at first; but at the end of a few months it was perceived that the mistake was distinctly the other way. The error is to be corrected. The town stands on high ground; it is about 700 feet above the sea level. It is so high that a wide view of river and lowland is offered from its streets. It is a very wonderful town indeed, and is not finished yet. All the streets are obstructed with building material, and this is being compacted into houses as fast as possible, to make room for more—for other people are anxious to build, as soon as they can get the use of the streets to pile up their bricks and stuff in. How solemn and beautiful the thought, that the earliest pioneer of civilization, the van-leader of civilization, is never the steamboat, never the railroad, never the Sabbath school, never the missionary—but always whisky! Such is the case. Look history over; you will see. The missionary comes after the whisky—I mean he arrives after the whisky has arrived; next comes the poor immigrant, with axe and hoe and rifle; next, the trader; next, the miscellaneous rush; next, the gambler, the desperado, the highwayman, and all the'r kindred in sin of both sexes; and next, the smart chap who has bought up an old grant that covers all the land; this brings the lawyer tribe; the vigilance committee brings the undertaker. All these interests bring the newspaper; the newspaper starts up politics and a railroad; all hands turn to and build a church and a jail—and behold, civilization is established forever in the land. But whisky, you see, was the van-leader in this beneficent work. It always is. It was like a foreigner—and excusable in a foreigner—to be ignorant of this great truth, and wander off into astronomy to borrow a symbol. But if he had been conversant with the facts, he would have said:

Westward the Jug of Empire takes its way.

"This great van-leader arrived upon the ground which St. Paul now occupies in June, 1837. Yes, at that date Pierre Parrant, a Canadian, built the first cabin, uncorked his jug, and began to sell whisky to the Indians. The result is before us."

A MINNESOTA LAKE.

Minneapolis.

This beautiful city is the twin sister of St. Paul, distant about ten miles westward on the Mississippi, although the suburbs of both are so rapidly nearing one another that the twain are likely soon to merge into a single metropolis. Thirty years ago the site upon which Minneapolis stands was a part of the Fort Snelling military reservation, and no foot save that of the red man had pressed the soil. At this point further navigation of the Mississippi River was barred by the Falls of St. Anthony, and one of the finest of water-powers awaited the genius of the Anglo-Saxon race to render it useful. To-day Minneapolis numbers at least 80,000 inhabitants, and the volume of her manufacturing and commercial business during the year 1882 fell little short of $175,000,000. Surely, even in this era of rapid growth in Western cities, to which the mind is somewhat accustomed, the development of Minneapolis into a place of its present importance is surprising.

The topography of the city is greatly in its favor. Situated on a broad plateau, high above the upper level of the river at the Falls, there is no danger from overflow; and yet the level of the place is so near that of the surrounding country, that the grades to and from the city admit the construction of rail and wagon roads with comparative ease, while the subsoil affords a foundation upon which the most massive buildings may be safely erected. The relation of Minneapolis to the surrounding country is everything that could be desired. The city lies on the eastern border of the great wheat belt of the Golden Northwest, and on the southern border of the pine and hard wood timber region of Minnesota. Here the wealth of raw material naturally finds its way to be conveniently converted into flour and lumber by the use of the grand water power, estimated at a capacity of 120,000 horses, within the city, and the product of the mills is afterwards forwarded to the markets of the world.

The Falls of St. Anthony, upon which the prosperity of Minneapolis is mainly founded, have a perpendicular height of eighteen feet, and the Mississippi has a rapid descent of eighty-two feet within the limits

Minneapolis, Minn.

N. LeVasseur
Quebec

Wm Watt Jr.
Brantford Ont

H. Hough,
 "World", Cobourg, Ont.
John J B Blink
 Belleville, Ont.

of the city. The view of the rapids above the cataract is very fine, but the picturesqueness of the water-fall has been sacrificed to purposes of utility. To prevent the wearing away of the ledge of rocks, a broad, smooth wooden apron has been constructed entirely across the river, sloping from the edge of the fall to a point far beyond its base, and, on reaching this, the water slips over, calmly and unvexed. The best view of the scene is from the magnificent suspension bridge of iron which spans the flood in graceful length, and with picturesque effect, at about the centre of the city. From this vantage-point an outlook is obtained upon the railroad tracks that stretch along below the bluffs, and also upon the river, with its channel above the falls almost choked with booms of logs that are to be cut into lumber by the extensive mills which line the shores. The water power is used for driving the machinery of the foundries, woolen mills and many other branches of mechanical industry of which Minneapolis is the seat.

Minneapolis is regularly laid out, broad avenues running from east to west, crossed by streets from north to south. The thoroughfares are usually eighty feet in width, with wide side-walks, shaded by rows of forest trees. There are many imposing business blocks, and the residence portion of the city is attractive, with its fine, spacious houses, and well kept lawns and grounds. On the outskirts of the city are thousands of pleasant cottages, which are the comfortable homes of the industrious mechanics who find employment in the mills and manufactories. Among the notable buildings are the Court House, City Hall, the Westminster Presbyterian and many other churches, the Nicollet House and First National Hotel. Other buildings in progress are, a new post-office, to be built by the Government at a cost of $175,000; an edifice for the Chamber of Commerce, $225,000; a Union Depot, $500,000; a new hotel, the "West," $1,000,000; and the Academy of Music, to be reconstructed at an outlay of $500,000. The amount expended upon building operations in the year 1882 was $8,375,000, being $3,310,000 more than were spent in the previous year. Indeed, so great is the demand for houses and stores in Minneapolis, that, as soon as the foundation of a structure is begun, the owners are beset with applications to hire. Real estate transactions to the extent of $18,701,256 were recorded for 1882.

There are seventy-one churches and missions in Minneapolis, each denomination being well represented. The Methodist lead the others with sixteen places of worship. Next come the Scandinavian Lutheran, with ten; the Baptist, with nine; the Episcopalian, with seven; the Congregational and Roman Catholic, each with six; and the Presbyterian, with five. The remaining twelve are divided among the minor sects. Some of the church edifices are of a high order of architecture.

The educational advantages supplied by the city are of a very high character. There are sixteen public schools, in which the standard of instruction is excellent. At the close of 1882 there were on the rolls about 8,000 pupils, under the charge of 184 teachers. The value of the school buildings is estimated at little short of half a million dollars. Supplementing the free schools is a free University. This is situated on the east side of the river, and is provided with substantial buildings and spacious grounds. The University of Minnesota is a State institution, founded eleven years ago, and so flourishing that it ranks among the leading educational establishments of the West. It was liberally endowed with lands by the United States Government, and the State itself steadily contributes to its current necessities. The curriculum is free to both sexes, and in 1882 there were 253 names enrolled in the usual classes, of which number seventy-two were female students. The work of the University is carried on by a large and efficient corps of instructors. It is, however, not confined to the ordinary scholastic courses, but includes also many branches of technical knowledge, giving especial attention to practical agriculture. The total number of students who attended the various classes and received free instruction during the year 1882 was 547. Connected with the institution is a well selected library of 15,000 volumes. The sphere of usefulness of the University is enlarging year by year, and the people of the State have set the seal of their approval upon what has been already accomplished. In addition to the public schools and the university, there are many private educational establishments of a high grade, among which are two seminaries for young ladies, and an efficient business college, besides thirteen denominational schools.

Minneapolis is well supplied with water from the river, and so abundantly that hundreds of the best residences have fountains playing day and night, keeping the lawns vividly green, and giving the entire city that beautiful vernal appearance which is the admiration of strangers. There were, in 1882, about twenty-five miles of water mains, distributing 8,500,000 gallons daily in summer. Ample provision is made to conquer fires, by a well equipped paid fire department. The streets are lighted by gas, and electric lights are also largely used. A sewerage system has been perfected ; horse-cars run through many of the streets, and a steam "dummy" engine drawing trains of open excursion cars brings the beautiful little lakes,

Cedar, Calhoun and Harriet (the two last named after the famous South Carolina senator and his wife), in the western suburb, within easy access of the heart of the city. The hotel accommodations are excellent; two theatres are well supported; charitable and benevolent organizations are generously sustained; there is a refined and cultivated society; several first-class newspapers, among which the *Tribune* takes rank as a metropolitan journal, besides other features of a large and prosperous city.

Exclusive of the Northern Pacific Railroad, several trunk lines enter Minneapolis, and make her the diverging point from which the country is penetrated in every direction. Among the other railroads which recognize the fact that Minneapolis is a great business centre, and compete for her trade, are the Chicago, Milwaukee and St. Paul, the St. Paul, Minneapolis and Manitoba, the Minneapolis and St. Louis, and the Chicago, St. Paul, Minneapolis and Duluth. These lines are constantly increasing their trackage, opening up new regions to settlement by running branch lines wherever the character of the country invites to railroad enterprise. Trains upon these roads, to the number of seventy-five or more, daily leave Minneapolis, some of which connect, without change of cars, with the entire railroad system of the United States and Canada. By the aid of the railroads Minneapolis feeds her mills and workshops, and distributes her merchandise, machinery and manufactures. At present the city is the theatre of extensive improvements, with the object of facilitating the enormous railway traffic. The Northern Pacific Railroad is about to construct large and convenient buildings for the accommodation of its great business, and purposes, also, to lay an additional track on the west side of the Mississippi River, from Anoka, twenty miles above, where the railroad will cross the stream. Among other improvements may also be named an immense viaduct, now nearly finished, and constructed at an outlay of $1,000,000, which will span the river, and be used by the trains of the St. Paul, Minneapolis and Manitoba Company. The Chicago, Milwaukee and St. Paul Railroad Company have recently erected car shops, covering five acres of ground, in the southern part of the city, and employing 1,500 workmen. Plans have been prepared for a union depot, which is to be one of the finest in the Northwest, and the construction of the building is to be pushed forward. In railway enterprises alone not less than $3,000,000 were to be expended in Minneapolis in the course of the year 1883.

The principal summer resort of Minneapolis is Lake Minnetonka, distant by rail fifteen miles from the city. "This," says a recent writer, "is one of the loveliest bits of water the tourist will find anywhere, for its depths have that deep azure tint that belongs to the purest water under summer skies, and its charmingly irregular shores, forested clear down to the shining beach, break into new combinations of woodland beauty at each advance of your boat. Upon the banks of this lake, and upon the islands that stud its bosom, many residences have been built, the summer homes not only of gentlemen who in the winter live in the neighboring cities, but also of many families from the South, even from New Orleans. This queen of the lake district is becoming more and more a favorite resort, and large preparations are making to accommodate summer visitors." There are many hotels on the lake shore, the largest of which, Hotel Lafayette, is capable of accommodating 1,200 guests.

The Mammoth Flour Mills.—It is aptly said that the history of the flour mills of Minneapolis is like the story of Aladdin. In 1860 the product was 30,000 barrels, and in 1882, 3,125,000 barrels. There are twenty-six mills in operation, the maximum daily capacity of all being 26,000 barrels. An idea of the gigantic proportions which this branch of industry has assumed may be obtained by remembering that the number of barrels of flour manufactured by one of the largest mills in the course of twenty-four hours is greater than that produced by an average-size mill in the course of a year. The capacity of the largest mill, the Pillsbury "A," is 5,200 barrels per diem; that of the Washburn "A," 3,000 barrels, and six other mills range from 1,200 to 2,000 barrels a day. The estimated quantity of wheat required to supply these mills in 1882 was 18,000,000 bushels. The capital invested in the flour milling industry is enormous, and the amount is constantly increasing. This is the result of the changes in the mode of manufacturing flour, which have been almost radical within the past few years. The use of the old mill stone has given place to the system of gradual reduction by iron rollers. The new process has not only raised the grade of flour, from the dark and inferior quality formerly produced, to the standard of the best Hungarian fancy brands, but has increased the quantity obtained from the grain as well as the capacity of the mills; thus better flour is now made at less expense than that which the inferior quality previously cost to manufacture. The flour of the Minnesota mills finds a ready market in all the Eastern cities, and also in Great Britain, France, Germany, Holland, Spain and Italy. Single orders are frequently taken for from 10,000 to 15,000 barrels, and the millers find it necessary, in securing the best trade, to control a great manufacturing capacity. Otherwise they would not be able to fill large orders promptly, nor obtain that uniformity in quality without which both the foreign and American market would soon be lost. Moreover, there is economy both in the construction and operation of a large mill over a small one. For example, the cost of one mill, with a capacity of 4,000 barrels daily, is much less than that of sixteen mills of 250 barrels capacity, or of eight mills of 500 barrels capacity, or even of four mills of 1,000 barrels capacity. The relative cost of operating a large mill is still less, and the chance of a uniform grade of flour is increased in the same ratio as the capacity of the mill. So medium-size mills, a few years ago considered the safest and most profitable, have been superseded by those of great capacity.

In order that some idea of a large Minneapolis flour mill may be obtained, the following facts relating to the Pillsbury "A" mill are given. This establishment is 180 feet in length by 115 in width, the building material being Trenton limestone, rock-faced, and laid in courses to the height of seven stories. Inside, on the basement floor, is a stone wall, 125 feet in length and fifteen in height, which holds the water from the canal after its passage from the falls before it descends to the wheels. Within this canal are the wheel-pits, dug out of the solid rock, fifty-three feet in depth. Inside these pits are flumes of boiler iron, twelve feet in diameter, in which two fifty-five inch wheels, each weighing, with the shafting, thirteen tons, are placed. The hydraulic power of a column of water, twelve feet in diameter, with a fall of fifty-three feet, is enormous. Only the strongest and toughest metal could withstand the strain. Seventeen thousand cubic feet of water rush down each flume every minute, and the combined force of the wheels is estimated at 2,400 horse-power, equivalent to that of twelve steam engines, each of 200 horse-power. This power is geared and harnessed to the machinery requisite to grinding 25,000 bushels of wheat in every twenty-four hours. On the first floor there are the main shafts of the driving apparatus, with pulleys twelve feet in diameter, weighing 13,000 pounds, over which runs belting of double thickness, forty-eight inches wide, at the rate of 4,260 feet in a minute. From the shafts also run thirty-inch belts perpendicularly to the attic floor, over eight-foot pulleys, at the rate of 2,664 feet per minute, furnishing the power which drives the bolting and elevating machinery. There are other pulleys and belting attached to the shafts for operating the rollers and purifiers, the electric light and other machinery. On this floor also is the wheat-bin for stowing grain. This holds 35,000 bushels, and extends through to the ceiling of the floor above, where it is connected with the weighing hopper. On the second floor the wheat is ground; the third floor is mainly devoted to packing; the fourth, fifth, sixth and seventh floors are filled with bolting chests.

middling-purifiers, bran-dusters and other machinery. Before going to the rollers to be ground into flour, the wheat is cleansed by passing through eight different sets of machinery. It is purged in this manner of wire, nails, cockle, small and imperfect kernels, and becomes actually polished before it is converted into flour. On the packing-floor the flour is discharged constantly from twenty-four spouts, and accumulates so fast that a car is either loaded with flour or bran every twenty-five minutes throughout the day. Any lack of transportation facilities at once clogs the mill. To every bushel of wheat there are thirteen pounds of bran or shorts; but for this "offal" there is a steady demand on the part of stock-raisers in the East. There are railroad tracks on either side of the mill, and the loading and unloading methods are complete. The establishment is provided with fire apparatus, electric lights, passenger elevator, machine shop, and every appliance for its convenient working. In fact, it is one of the model flouring mills of Minneapolis, and the visitor who examines its features in detail will be well repaid.

The process of manufacturing flour in a typical Minneapolis mill are clearly described by Ernest Ingersoll, in *Harper's Magazine* for June, 1883:

"When the wheat comes in it is unloaded from the cars, by the aid of steam shovels, into a hopper bin, whence it is elevated to the fifth floor, and fed into a receiving bin, the bottom of which extends down to the fourth floor. Out of this it empties itself into conveyers, consisting of small buckets, traveling upon an endless belt, and is taken to storage bins on the first and second floors. Here it rests until wanted for milling. When this time comes the wheat travels by conveyers to the top floor, whence it is fed down into the grain separators in the story beneath, which sift out the chaff, straw, and other foreign matter. This done, it descends another story upon patented grading screens, which sort out the larger-sized grains from the smaller, the latter falling through the meshes of the screen, after which the selected portion drops into the cockles on the floor beneath, and, these escaped, falls still further into the Brush machines. All this time the wheat remains wheat—the kernel is entire. Its next move, however, begins its destruction, for now the ending stones are encountered, which break the germinal point off each grain. This matter accomplished, the wheat is shot away up to the attic again, and, traversing the whole length of the mill, falls into an aspirator on the seventh floor, having passed which, it slides down to the second floor, and is sent through the corrugated rollers. These rollers have shallow grooves cut spirally upon them, with rounded ridges between. The opposing rollers are grooved in an opposite direction, and it is impossible for a grain of wheat to get through without being cracked in two, though the rollers are not sufficiently near together to do much more than that. It comes out of this ordeal looking as though mice had chewed it, and, pouring into special conveyers, speedily finds itself up on the seventh floor again, where the flour dust which has been produced by this rough handling is bolted out in reels, and all that is left—no longer ***wheat***—is divided into 'middlings' and 'tailings.' The tailings consist of the hard seed case and the refuse part, and go into market as 'feed' and 'bran,' while the middlings are reserved for further perfection into flour; they are the starchy, good centres of the grains.

"The first operation toward this end is the grading of the middlings, for which purpose they pass upon silken sieves arranged in narrow horizontal troughs, and given a gentle shaking motion by machinery. There is a succession of these bolting cloths, so that the middlings pass through ten gradings. Next, they go to a series of purifiers, which resemble fanning machines, and thence to corrugated rollers, each successive set of which are more closely apposed, where the meal is ground finer and finer. There are five of these corrugations in all, and between each occurs a process of bolting to get rid of the waste, and a journey from bottom to top of the mill and back again. Nevertheless, in spite of all this bolting, there remains a large quantity of dust, which must be removed in order to make the flour of the best quality. And hereby hangs a tale of considerable interest to Minneapolis men.

"In the old mill which not long ago occupied the site of this new one there stood upon one side the usual rows of buhrs, in this case twenty in number. Through the conveyer boxes connected with them was drawn a strong current of air that took up all the fine particles of flour dust, and wafted it with the strength of a tempest into two dust rooms, where it was allowed to settle. The daily deposit was about three thousand pounds, which was removed every morning. In addition to these small chambers there were several purifiers on the upper floors, that discharged their dust right out into the room. The atmosphere of the whole mill thus became surcharged with exceedingly minute and fuzzy particles, which are very inflammable, and, when mixed in certain proportions with the air, highly explosive. This mixture had apparently been brought by the millers to just about the right point, when fate supplied a torch. A piece of wire fell between the buhr stones, or into some rollers, and began a lightning express journey through the machinery, in the course of which it became red hot, when it found an exit, and plunged out into the air. It was a most startling instance of the conversion of heat into motion. A lighted match in a keg of powder is the only analogy to illustrate the result. One room down-stairs burst into flames, and the watchman had only time to pull the electric fire alarm near his hand, when he and the mill together disappeared from the face of the earth. A terrific explosion, generated throughout that great factory in an instant, rent all parts of the immense structure as suddenly as a child knocks over a tower of cards, leaving nothing but blazing ruins to show where, a twinkling before, had stood the largest flour mill in the country. Nor was this all. The land was dug from under the foundations, and the massive machinery buried out of sight. Two other mills and an elevator near by were demolished, so that not one stone remained above another, while of three other mills, cracked and tottering walls, and charred interiors, were the only mementoes of the day's flourishing business.

"The good that came out of this seemingly wholly harmful episode, which scratched an end mark to one era of the city's prosperity, was the introduction into the new mills of a system of dust-saving that renders such a calamity improbable, if not impossible, in future. Now, instead of being thrown abroad into a large room, the dust is discharged by suction pans into close, fire-proof receivers, where it accumulates in great quantities, and is sold as a low grade of flour. This dust having been removed, what remains is the best quality of flour. It is barreled by the aid of a machine permitting the precise weight of 196 pounds to be determined, packed and branded with great speed.

"Bakers, however, use what is known as 'wheat' or 'straight' flour, which is the product of the five reductions, all the subsequent processes through which the middlings pass in making fine flour being omitted. 'Fancy' flour differs from the ordinary superfine in that the middlings are ground through smooth rollers."

Mark Twain, in his book before quoted, devotes a paragraph to Minneapolis as follows, but the figures which are given have been superseded by much larger ones:—

"All that I have said of the newness, briskness, swift progress, wealth, intelligence, fine and substantial architecture, and general slash and go and energy of St. Paul, will apply to his near neighbor, Minneapolis—with the addition that the latter is the bigger of the two cities. These extraordinary towns were ten miles apart a few months ago, but were growing so fast that they may possibly be joined now, and getting along under a single mayor. At any rate, within five years from now there will be at least such a substantial ligament of buildings stretching between them, and uniting them, that a stranger will not be able to tell where the one Siamese twin leaves off and the other begins. Combined, they will then number a population of 250,000, if they continue to grow as they are now growing. Thus, this centre of population at the head of the Mississippi navigation, will then begin a rivalry as to numbers with that centre of population at the foot of it— New Orleans. Minneapolis is situated at the Falls of St. Anthony, which stretch across the river, 1,500 feet, and have a fall of eighty-two feet—a water power which, by art, has been made of inestimable value, business-wise, though somewhat to the damage of the falls as a spectacle, or as a background against which to get your photograph taken. Thirty flouring mills turn out 2,000,000 barrels of the very choicest of flour every year; twenty saw mills produce 200,000,000 feet of lumber annually; then there are woolen mills, cotton mills, paper and oil mills; and sash, nail, furniture, barrel and other factories, without number, so to speak. The great flouring mills here and at St. Paul use the 'new process,' and mash the wheat by rolling, instead of grinding it. Sixteen railroads meet at Minneapolis, and sixty-five passenger trains arrive and depart daily. In this place, as in St. Paul, journalism thrives. Here there are three great dailies, ten weeklies, and three monthlies. There is a University, with 400 students—and, better still, its good efforts are not confined to enlightening the one sex. There are sixteen public schools with buildings which cost $500,000; there are 6,000 pupils and 128 teachers. There are also seventy churches existing, and a lot more projected. The banks aggregate a capital of $3,000,000, and the wholesale jobbing trade of the town amounts to $50,000,000."

St. Paul Division.

ST. PAUL TO BRAINERD.—DISTANCE, 136 MILES.

From St. Paul to Sauk Rapids, a distance of seventy-five miles, the Northern Pacific Railroad Company is at present using the tracks of the St. Paul, Minneapolis and Manitoba Railroad, under a lease. But the Northern Pacific Railroad is extending its own line on this important branch, and proposes to have double tracks between Sauk Rapids and Minneapolis at an early day.

After leaving Minneapolis, the course of the railroad, following the left hand shore of the Mississippi River, is somewhat west of north, the route being through a level, or gently undulating, region. The surface of the country is, however, diversified by lakes, rivers, and small tracts of prairie and growths of hard wood timber. Passing the unimportant stations of *Fridley* and *Coon Creek*, the first place of any prominence after leaving Minneapolis is

Anoka.—This town, the county-seat of the county of the same name, is situated twenty miles from Minneapolis, at the mouth of Rum River. It contained at a recent date at least 4,000 inhabitants, who carry on a flourishing trade in lumber and flour, as well as in a variety of other manufactures. Anoka has several churches, excellent schools, enterprising newspapers, good hotels, and the usual stores for supplying the necessities of a thriving town with outlying lumber camps and farming communities.

Six miles beyond Anoka is *Itasca*, and opposite this small station, at the mouth of the Crow River, is the thriving town of *Dayton*. *Elk River Station*, five miles from Itasca, is the stopping-place for *Orona*, distant a mile and a half from the railroad, but not visible through the forest. Orona is the seat of large milling operations. A stage leaves Elk River Station three times a week for *Princeton*, nineteen miles to the northward, which is the lively head-quarters of the lumbermen of the Upper Rum River. *Big Lake Station*, nine miles farther, deriving its name from the beautiful sheet of water on which it is situated, is the nearest point to the German and Swedish settlements near *Eagle Lake*, some miles distant. *Becker* and *C'ear Lake Stations*, the next halting places, afford outlets to a rich farming and grazing region, which stretches away on both sides of the Mississippi, for many miles of its course, and whence large quantities of wheat and dairy products are shipped to the markets of Minneapolis and St. Paul. The populous agricultural and milling towns of *Groton*, *Monticello*, *Clearwater*, *Buffalo*, county-seat of Wright County, and *Fairhaven*, are embraced in the area tributary to the railroad stations named.

At *East St. Cloud Station*, sixty-three miles from Minneapolis, a fine wagon bridge spans the Mississippi, leading directly to the large and flourishing city of *St. Cloud*, a place of nearly 4,000 inhabitants. Near the station are extensive granite quarries, which produce an almost unlimited supply of the finest building stone. Two miles beyond is the town of

Sauk Rapids—(75 miles northwest from St. Paul; population, 1,200.)—The village of Sauk Rapids, the county-seat of Benton County, lies on the east bank of the Mississippi River, at the falls of Sauk Rapids, from which its name is derived. The Mississippi River at this point is 600 feet wide, and has a fall of eighteen feet in one mile. The place is of much importance on account of the extensive beds of granite in its immediate vicinity, the stone, it is said, being equal to the celebrated Quincy granite of New England, varying only in color, and also for the reason that a fine water power is furnished by the rapids which begin where the Sauk River enters from the west, at the upper end of the village. The rapids continue over a bed of granite a distance of half a mile, and, viewed from either bank, present a picture of great beauty. A substantial bridge, crossing the river just below the main falls, affords a fine view of the scenery above and below. The village has a graded school, Episcopal, Congregational and Methodist churches, four hotels, two flouring mills, with a capacity of 500 barrels daily, a saw mill, four general stores, a large hardware store and work shops. Sauk Rapids occupies a slightly elevated position, bounded on the north and east by low wooded bluffs, which form the abutment of the higher level of the surrounding country. The large expenditure of capital necessary to utilize the water power here has retarded the growth of the place; but the construction of railroads to various points gives increased facilities for the exchange of manufactures, and the extensive improvement of the country around, in the opening up of farms by incoming settlers, renders the use of the power, to a far greater extent than now, a necessity of the not distant future.

Persons in pursuit of pleasure are here amply able to indulge their tastes. Fine streams and lakes abound in the vicinity, and in all of them there is excellent fishing. Deer, ducks and other small game are plentiful, and an occasional bear offers fine sport. Passing the unimportant station of *Watab*, seven miles beyond Sauk Rapids, the next stopping place is

Rice's—(88 miles from St. Paul; population, 550.)—Rice's is situated in Benton County, one and a half miles east of the Mississippi River, and two miles north of Little Rock Lake. It is in the midst of a good farming community, and is the point of departure for the Rum River lumber regions. It has five general stores and two good hotels, capable of accommodating 100 guests each. The place increased in population nearly one-third during 1882, and is growing steadily. It was named after George T. Rice, one of the pioneers, who, for many years before the era of railroads, kept a tavern and stage station half a mile from the present town site. Lumber and wheat are the principal exports. There is good hunting and fishing in the neighborhood.

Royalton—(95 miles north of St. Paul; population, 200.)—Royalton was founded in 1880. It is situated two miles east of the Mississippi River, in the midst of a prairie, dotted with groves of hard wood trees. The population, composed of Americans, Germans and Swedes, is enterprising, and the town is growing fast. There are an Episcopal and a Presbyterian church; a good school; a saw mill with a capacity of 20,000 feet per day, which may be easily increased; two hotels; two grain warehouses, each capable of storing 8,000 bushels; a steam elevator, close to the railroad track, with a capacity of 30,000 bushels; a livery stable, and numerous stores and shops. Two excellent mill sites, well adapted for the manufacture of flour, are offered by the Platte River, a pleasant stream which skirts the town. A large amount of wheat is annually shipped from the station by the surrounding farming population. There are three lumber settlements on the west side of the Mississippi, in the heart of the hard wood timber region, within six miles of Royalton, named North Prairie, Two Rivers and Elmdale. These places are tributary to Royalton, employing hundreds of teams in hauling logs, cord wood and railroad ties to the station. Every year thousands of cords of hard wood, principally maple, are shipped, and the railroad draws largely upon the neighborhood for ties, thousands of which are piled up on either side of the track. As showing the extent of lumbering operations in this part of Minnesota, it is said that in the summer of 1882 not less than 170,000,000 feet of logs passed by Royalton, floating down the Mississippi and Platte Rivers to the saw mills at various points below. The view on approaching the town is quite pleasing. On the west are broad stretches of prairie between the railroad and the river, across which may be seen the settlement of North Prairie, while on the other side is a line of pine forest, broken here and there by a farm clearing and the homes of prosperous settlers. Hunting and shooting in this vicinity present their attractions. Deer are plentiful, and feathered game so abundant that prairie chickens, wild geese, ducks and grouse are shipped in quantities during the season to the markets of St. Paul and Minneapolis. *Portland*, five miles beyond, is simply a side track.

Little Falls.—(105 miles northwest of St. Paul; population, 1,100.)— This town is nicely situated on the east bank of the Mississippi, thirty-one miles south of Brainerd. It has two newspapers, good hotels, schools, churches and general stores. It is the junction of the Little Falls and Dakota Division of the N. P. R. R. with the W. R. R. Division of the same road, and is also a depot for the various lumbering firms who have camps in the neighborhood. The town is in Morrison County, and takes its name from the falls at this point in the Mississippi River, which, if utilized, would furnish water power only second to that in Minneapolis. The scenery is diversified and interesting. Finely wooded bluffs cropping up between rich prairies make the neighborhood favorable for hunting. Five miles east is a pleasant inland lake named *Rice Lake*, from the large quantity of wild rice growing around its shores. This is a resort for wild ducks, and in season large numbers are bagged. The woods abound with partridge, and the prairies with grouse or prairie chickens, while deer are found in great numbers within easy distance. Over thirty tons of saddles of venison were shipped from this point during the season of 1882. It is principally a lumber and farming district, shipping large quantities of vegetables to St. Paul and Minneapolis.

A branch road leaves Little Falls, running in a course slightly south of west through a beautiful country, well wooded and watered, and finely adapted to mixed farming, a distance of eighty-eight miles to Morris, Minn. This branch is called the Little Falls and Dakota Railroad, and its construction will be pushed to the fat prairie lands of central Dakota.

Little Falls and Dakota Branch.

FROM LITTLE FALLS TO MORRIS.—DISTANCE, 88 MILES.

La Fond—(7 miles from Little Falls; population, 60.)—This place contains a saw mill and a general store. The principal industry is lumbering, and shipments are made of lumber, wood, posts, piling, etc.

Swanville—(16 miles from Little Falls; population, 120.)—Swanville has one hotel, general stores in all branches of trade, elevator, and a saw mill. This is a thickly timbered region, and also well adapted to grazing.

Grey Eagle—(26 miles from Little Falls; population, 260.)—This town, although little more than a year old, shows signs of developing into a lively business place. There are two hotels, one saw mill, two general stores, a church and a school-house. The country is well wooded with fine hard wood, and the chief industry is the marketing of wood, ties and lumber. The woods are well stocked with game, and the many beautiful lakes with fish. *Twin Lakes*, just east of the town, are situated near the railroad, and are a favorite resort for hunters and fishermen. *Mound Lake*, three miles northeast of the town, has a high, firm beach. There is a mound upon the north side, the top of which is 180 feet above the surface of the lake. From the apex of this mound, a commanding view is afforded of Little Falls, thirty miles away, and several towns in other directions. Old fishermen say this is the best lake in Minnesota for fishing. *Birch Lake*, lying one mile west and half a mile south of the town, is noted for its fine gravel beach, and is considered the finest pleasure resort in the surrounding country. Upon the shores of the lake an hotel has been built to accommodate the pleasure seekers, who go there to spend the summer months, or to enjoy the fine hunting and fishing found in the vicinity. In Birch Lake fish of from ten to twenty-five pounds weight are often caught.

Birch Bark—(29 miles from Little Falls; population, 100.)—Birch Bark has one good hotel, three stores and a saw mill. The shipments are lumber. The surrounding country is heavily timbered, and also affords good pasturage.

Sauk Centre—(37 miles from Little Falls; population, 2,000.)—This town is situated on Sauk River at the outlet of Sauk Lake. It forms the natural geographical and business centre and outlet of an extensive area of rich agricultural country, well supplied with timber and water, and finely adapted to raising grain and stock, as well as to dairy purposes. The country south and west is rolling prairie, interspersed with groves and lakes. On the north and east there is principally timber land. The first settlement on the present site of the town was made in 1856 by the "Sauk Centre Town Site Co.," composed of seven persons. The place is in a flourishing condition, and its enterprising citizens are building up a substantial business and commercial standing. There are Episcopal, Congregational, Methodist, Baptist, Roman Catholic and Lutheran churches. A fine brick school building has just been finished, in which eleven teachers are employed. Sauk Centre has two newspapers, two large flouring mills, two banks, dry goods, clothing, drug, grocery, crockery and hardware stores, machine shops, manufactories of agricultural implements, and also a water power that might be used far more extensively. The railroad crosses an arm of Sauk Lake, but only a small part of it can be seen from the cars. The lake is twelve miles in length, and is partially hidden by a point of land projecting into it. There is very good pickerel, bass and perch fishing. Prairie chickens are found only a short distance from town, and, in season, ducks and geese come by thousands to the marshes west of and along the line of the railroad. Many deer and wild cats, and occasionally a bear and panther, frequent the heavy timber to the north and northwest. The remains of an old stockade, built by the early settlers for protection against the Indians, are still standing. The Big Woods, which lie directly north of Sauk Centre, are famous for their past history, having long been the bourne upon which the eyes of the Dakota and Ojibway tribes were fixed, and for which battles without number were fought, and every inch of the ground contested. This region was at first held by the Dakotas, but the more subtle cunning of the Ojibways finally prevailed, and they came into possession of it. Before the ascendency of these two tribes, wandering bands of the Alabaska nation alternately had possession. The heavy forests, winding streams, beautiful prairies and meadows, being the best hunting grounds in the State, were dear to the savage heart.

Westport—(48 miles from Little Falls; population, 50)—is quite a new place, containing an elevator for the storage and shipment of wheat. The soil of the surrounding country is rich and fertile, adapted to raising all kinds of farm products.

Villard—(53 miles from Little Falls, and 116 miles from Minneapolis; population, 450.)—This village, named after the President of the Northern Pacific Railroad, is situated in Pope County. A year ago, the present site of the town was a wheat field, and afforded a camping place for a party of sixty gentlemen from England, who came annually to enjoy themselves in bagging game and catching fish at a chain of pretty lakes beginning one-third of a mile from the village. The first train reached Villard about September 1st, 1882. On the 13th of the same month several car loads of lumber arrived, and an enterprising man began the erection of a store. This was the signal for a rapid immigration, lots speedily advancing from $50 to $200. The growth has been steady ever since, there being already established Baptist, Methodist and United Brethren church societies, with schools, the usual stores, and all other appurtenances of a thriving village. The soil of the surrounding country is a black loam with clay subsoil. Farmers bring their wheat twelve miles to the village for shipment, and large quantities are forwarded to market. The scenery about Villard is particularly delightful. The chain of lakes near the village are: Lake Villard, one and three-fourths of a mile north and south, and one and one-fourth of a mile wide; Lake Amelia, four miles long by one mile wide; Lake Levan, two miles long by one and one-half miles wide; Lake Ellen, one and one-half miles long by one and one-third miles wide. These lakes are all connected. Their banks are lined with maple and oak of heavy growth; the waters are pure and clear, and the bottoms gravelly. They all abound with black bass, which makes them one of the best fishing grounds in the entire section. There are plenty of water fowl, and in fact there is no end of amusement for the sportsman. Between two of the lakes, for a distance of nearly one mile, the railroad track runs, there being just room enough for a single track—the road-bed sloping down on either side to the water. From the car windows the view makes a very pretty picture. The village of Villard, from the nature of its surroundings is likely to become one of the greatest summer resorts in the Golden Northwest. Already preparations have been made to receive tourists.

Glenwood—(60 miles from Little Falls; population, 500.)—Glenwood, the county-seat of Pope County, is situated in a small circular valley at the eastern extremity of Lake Minnewaska, or Lake Whipple, as it has sometimes been called. The hills on the north and east rise 280 feet above the level of the lake, and a little above the surrounding prairie. These eminences are cut up with deep wooded dells or ravines, through which flow clear creeks of spring water, pursuing their way across the valley to the lake below. The railroad station is prominently situated 200 feet above the valley, and offers a pretty view of the village, with its substantial brick court-house, school building and church, its neat residences, and the silvery lake, whose shores are fringed with oak and maple. Lake Minnewaska, now in sight from the cars between Glenwood and Starbuck, and anon hid from view by hills or groves of timber, has a clean, gravelly beach, which affords a delightful drive, over six miles in length, shaded by a growth of forest trees. The lake abounds in fresh water fish of all kinds. Many times in spring wagon loads are taken away, the result of a night or day's fishing. Many springs gush from the hill slopes on the north shore of the lake, within half a mile to two miles from its eastern end. Some of these springs are pure, sweet water, while others contain iron and potassium, or are strongly impregnated with sulphur. One of these fountains, coming out of the bluff just behind the village, forms a considerable stream, which has been dammed by the owner, and made to furnish the power to operate his mill. Of the excellent medicinal properties of the mineral springs there is ample evidence, as invalids using the water will testify.

When first Glenwood was settled, in 1866, there was but one building, 10x12 feet in size. Up to the advent of the Little Falls and Dakota Railroad the place grew slowly, but during 1882 the population nearly doubled. Glenwood has two hotels, three stores, two newspapers and a brick yard.

Starbuck—(69 miles from Little Falls; population, 150.)—The town of Starbuck, situated on the western end of Lake Whipple, at present comprises about 100 buildings, all erected after the railroad reached the place in the autumn of 1882. Among them are an hotel, a drug store, a hardware store and five general stores. The town will doubtless have a large growth, as elevators, a second hotel and stores of various kinds are in process of erection. *Lake Whipple* was named in honor of Bishop Whipple, of St. Paul, who spends a long vacation each summer in its vicinity. It is ten miles long by three miles in width, and has a sandy beach, which is partially skirted with timber. The fishing is of the finest in the State, pike, pickerel, red-horse, buffalo, bass and perch abounding in large numbers. Flocks of ducks and geese offer an excellent opportunity to the sportsman, as well as prairie chickens in their season. Foxes and wolves are also to be found in the vicinity. The lake is fed by numerous springs, which are noted for their medicinal qualities.

Scandiaville—(79 miles from Little Falls; population, 100.)—This place, growing rapidly, has one hotel, an elevator and general stores. The country is well suited to farming, and an abundance of small game is found in the vicinity.

Morris—(88 miles from Little Falls; population, 1,500.)—Morris, the county-seat of Stevens County, is the present western terminus of the Little Falls and Dakota branch. The town has four churches, a graded school, a public hall, two banks, three good hotels, one newspaper, two flouring mills, three elevators and a large number of business houses. It is in a prosperous condition, and is fast becoming an important city. A good quality of cream-colored brick, manufactured here, is used in the construction of the buildings. A new court-house with jail has been erected, to cost $20,000. A large creamery has also been established. The chief industries of Morris are stock raising and agriculture. The shipments are wheat, barley, oats and corn. There are several lakes well stocked with fish in this vicinity, and flocks of prairie chickens, snipe, plover and ducks are found here in their season.

St. Paul Division.

[*Continued from page 38.*]

Belle Prairie—(109 miles from St. Paul; population, 800.)—This town, in Morrison County, four and a half miles north of Little Falls, on the east bank of the Mississippi, derives its name from the beautiful, level strip of prairie, about twelve miles long, and varying from two to four miles in width, upon the edge of which it is situated. Further eastward the surface is more rolling, and partly bordered by wooded bluffs, which finally merge into a region heavily timbered, chiefly by different varieties of hard wood. Looking westward, the prairie is skirted by a strip of timber which obstructs the view of the Mississippi. West of the river the surface is undulating—prairie and timber alternating with wild meadows and swamps. The soil of Belle Prairie is a rich, black sand, and well adapted to all kinds of agricultural products, especially wheat, potatoes and garden vegetables. About 7,000 acres are under cultivation. The population of the country tributary to Belle Prairie is 1,000, the majority being French Canadians, who are mostly engaged in agricultural pursuits and lumbering.

There are here an hotel, a post-office, stores and shops, an elevator, a public hall, district schools, a Roman Catholic church, and a convent, with which a school, attended by fifty scholars, is connected. This town is one of the oldest settlements in northern Minnesota. Mr. Frederick Ayer, a missionary, settled here in 1818, and erected a commodious school-house for the education of Indian children. The school was for a time in a flourishing condition, and employed a number of teachers; but, after the removal of the Indians, the enterprise was abandoned. Game, comprising deer, ducks and prairie chickens, is abundant.

Fort Ripley—(119 miles from St. Paul; population, 100.)—This station derives its name from the now unoccupied fort, distant one mile, on the west bank of the Mississippi River. It is the shipping point of a rich lumbering and farming region, into which settlers are rapidly entering. *Albion* (five miles beyond) is only a siding.

Crow-Wing—(128 miles from St. Paul.)—is an unimportant station, but is associated with the Indian history of the region. The venerable Bishop Whipple, of Minnesota, who is full of memories of the stirring pioneer days of the country, narrates the following anecdotes, which show that the red men are not devoid of wit. The celebrated Hole-in-the-day, head chief of the Chippeways, had a house a little west of the station. It was a palatial residence for the Indian country, and Hole-in-the-day put on all the airs of an Eastern prince. He was a fine-looking specimen of his race; straight as an arrow, with clear, dark olive complexion; bright, flashing black eyes, and as scrupulously neat in dress as a Broadway dandy. He was cunning, unscrupulous, and suspected by the Indians as winking at the corrupt practices of the agency in the old time. On one occasion, the Government sent out an inspector to investigate the agency. The agent purchased Hole-in-the-day a beautiful bridle and saddle a few days before the inspector came. When the inspector called on Hole-in-the-day, the Chief showed him among other treasures this bridle. Putting the bit in his mouth, he said with a laugh: "This no horse bridle; this Indian bridle. Indian no talk." Hole-in-the-day was murdered by some of his own tribe about eleven years ago. He had aroused the anger of the adherents of "Little Crow," the instigator and leader in the horrible massacre of over 600 settlers in Minnesota in 1862, by turning Government informer. Out of this grew the unfriendly feeling which resulted in his death. On another occasion, a special agent came out, who thought he could awe the Indians by a grand speech. The department had prepared a treaty, by which the Gull Lake, Mille Lac, Rabbit Lake and Sandy Lake Indians were to sell their lands, and accept a home in a desolate country, north of Leech Lake. The agent rose and said: "Your great father has heard of your wrongs, and determined that he would send you an honest man. He looked north, east, west and south, and when he saw me, he said: 'I have found an honest man; go and make a treaty with my red children.' Look at me: the winds of fifty years have blown over my head and silvered it over with gray. In all that time I have never done wrong to any man. I tell you as a man, who never tells lies, it is wise for you to sign this treaty at once." He sat down amid breathless silence. Shah-bah-skong, a chief of the Mille Lac Indians, rose and said: "My father, look on me. The winds of fifty years have blown over my head, and silvered it over with gray. The winds have not blown my brains away." There was a shout of laughter, and the council ended.

Brainerd —(136 miles from St. Paul, and 114 miles from Duluth; population, 7,000.)—Brainerd, City of the Pines, is situated on the east bank of the Mississippi River, on the main line of the Northern Pacific Railroad, at an elevation of 1,600 feet above the sea. It embraces within its limits an area of about 1,500 acres, and is platted parallel to and at right angles with the railroad on either side, in the midst of the pine forest, being one of the most important, picturesque and attractive towns on the line of the railroad, north of Minneapolis, and west of the great lakes, in Minnesota. There are seven church organizations, each worshiping in its own edifice; three school buildings, a court-house and jail, hotels, two daily and two weekly newspapers, and a public hall. Approaching the town from the south and east, the eye is attracted by the lofty smoke-stack (110 feet high) of the Railroad Company's shops, which here cover an area of about twenty acres, and consist of round-house, containing forty-four stalls; machine shop, with capacity for handling twenty-two locomotives at once; boiler house, copper shop, blacksmith forges, foundry and numerous other accessories of the head-quarters of the motive power of a great railroad. Passing by this busy hive of industry, going west, the traveler is at once ushered into the business portion of the city, which stretches along parallel to the track on the south side for a distance of nearly half a mile. Conspicuous here is the Sleeper block of stores, and the First National Bank block, erected in 1882, at a cost of $25,000. On the north side of the track are obtained glimpses, through the timber, of picturesque residences, the Episcopal and Congregational churches, Gregory Park, enclosing ten acres of stately pines, and the court-house and jail, erected at a cost of $30,000. Here also is the imposing building belonging to the Railroad Company, and occupied as the head-quarters of division offices. The Brainerd Water and Power Company, with a paid up capital of $100,000, supplies the city with water from the Mississippi River, and also contemplates furnishing the city and shops with the electric light. An hotel, containing eighty rooms, with all modern improvements, has recently been built. There is here a steam saw-mill, employing a capital of $200,000, whose output of lumber in 1882 was 16,000,000 feet, and 25,000,000 shingles and lath; a brick yard, with a plant valued at $30,000, which has large contracts already in hand for bricks for local use alone. The Railroad Company has rebuilt its general hospital for the use of its employees who may be disabled by sickness or accident, and plans for other extensive improvements by the Railroad Company are in a forward state of preparation. It is, therefore, not improbable that the disbursement of currency will hereafter exceed $75,000 per month at this station.

Brainerd is the gateway to the vast lumber region north and east to the sources of the Mississippi. Good wagon roads penetrate the forest in all directions, and a stage line and semi-weekly mail service is maintained to *Leech Lake* and *Lake Winnebagoshish*, which the United States Government is converting into huge reservoirs, at an expense of half a million dollars, to regulate the stage of water in the upper Mississippi. Leech Lake contains an area of 200 square miles, and Winnebagoshish half as much more. During the season of navigation a small steam-boat plies between Aiken and Pokegama Falls, where the Mississippi takes a sudden leap of seventeen and a half feet, around which is a short "carry," or portage, whence a small Government steamer penetrates to the Government works above. An hundred lakes, at varying distances of three to twenty-five miles from Brainerd, and of easy access, are stocked with black bass, wall-eyed pike, pickerel, muskelonge and other varieties of fish, all of exquisite flavor; numerous rice lakes afford breeding places for myriads of water fowl, while the forest is full of game and fur-bearing animals. Red deer and pheasants may be taken by the sportsman, within easy strolling distance of the town, and a black bear, wolf or wolverine often add piquancy to the hunter's quest. There is an hotel at *Gull Lake*, twelve miles distant northwest, with accommodations for twenty guests, and at *Serpent Lake*, sixteen miles northeast, there are accommodations for perhaps an equal number. *Mille Lac Lake*, twenty-two miles southeast, is the largest, and perhaps the most charming, of all the Minnesota lakes. Embowered in a magnificent forest of butternut, ash, sugar maple and other hard woods, its solitude has rarely been disturbed by the sound of the woodman's axe. It has an area of nearly 400 square miles, and a gravelly beach skirts its shores for nearly 100 miles. This lake is the source of the Rum River; its waters teem with fish, many of which are of marvelous size; black bass, of ten and twelve pounds each, are often hooked. Its shores abound with game, attracted hither in the fall by the immense crops of mast in the forest, and wild rice in the thousand lakes. Openings in the forest, bits of prairie and meadow, produce wild strawberries, blueberries, raspberries and cranberries, hundreds of bushels of which are annually shipped from this station; the undergrowth is rich with ferns, and flowers and flowering shrubs of exquisite beauty. Active measures are in progress for making Mille Lac, with its manifold attractions, available to the public. A town site has been selected, surveyed and platted on the west bank of the lake, on a plateau elevated some fifteen or twenty feet above its level, and sloping gently toward it, which has been named Rowe, in compliment to the editor of the *American Field*. Its proprietors are perfecting plans for its improvement by the erection of an hotel, and have already completed arrangements for putting on the lake a number of pleasure boats and a steam launch.

A Chief's Stratagem.— During the Sioux outbreak in 1862, Hole-in-the-day, and some kindred spirits, sought to excite the Chippeways to hostility. They sent runners to Leech Lake, sixty miles north of Brainerd, with directions to seize and kill the traders. The Indians captured Messrs. Rutherford, Sutherland, Whitehead and others. The war party were in the ascendant; but an old chief, "Buffalo," was the friend of the prisoners. He knew that pleas for mercy would be unheeded. He went to the council, and rose and said: "My children, the whites have wronged us; their children ought to die. But there is a right and a wrong time to kill men. Our chiefs at Crow Wing have sent us word that there is a war between the red men and white men. We are far away in the woods. If, when we get to Crow Wing, there is no war, and our great father looks in our faces and says, 'Where are my children?' we shall look foolish. Some of us will be hanged. We will not kill the white men now; we will keep them to be killed by and by." All the Indians shouted, "Ho! ho! good, good!" A few days after they were all released by Buffalo, and owed their lives to this good old chief.

The Indians of Leech Lake bear the name of Pillagers. Years ago the factors of the Northwest Fur Company were wont to leave their stores unlocked, and property was more safe than among civilized men. Hon. H.

M. Rice, the life-long friend of the Chippeways, whom they call "Wabahmanomin" (white rice), bears testimony to their fidelity, honesty and devotion. At the time Gen. H. H. Sibley was chief factor among the Sioux, it was their boast that they had never taken the life of a white man, and during thirty years he only knew of one theft from his trading post.

The early traders were always the Indian's friend. The tie which bound them together was one of mutual dependence. If the trader oppressed the Indians, they would not pay him their debt; if they did not pay, he could not remain. There never were more lasting friendships than between the old traders and their Indian friends.

Origin of Mother Goose.—Travelers, as they pass through the Indian country, notice that the Indian mother straps her babe upon a board which has a soft covering of down or hair. Over the child's face is a hoop by which the board is lifted, and, when the mother is picking berries, she hangs this cradle on the limb of a tree. You see it, and for the first time understand Mother Goose.

> "Rock-a-by, baby, in the tree top,
> When the wind blows the cradle will rock,
> When the bough breaks the cradle will fall,
> Down comes baby, cradle and all."

Wisconsin Division.

FROM SUPERIOR TO THOMPSON.—DISTANCE, 22 MILES.

The Wisconsin Division of the Northern Pacific Railroad extends from Montreal River, the western boundary of Michigan, across the State of Wisconsin, *via* Superior, at the head of Lake Superior, and thence to Thompson, at which point it connects with the road from Duluth. The length of this division is 120 miles, and the road will be built throughout the entire route as soon as practicable. The portion of this division of the railroad now in operation extends from Superior to Thompson, a distance of twenty-two miles. Between these points the railroad passes through heavy forests of pine, and opens the way also to the extensive lumber regions of northern Wisconsin. On this division of the Northern Pacific Railroad, during the next few years, the largest lumber mills in the Northwest will be established.

Superior—(Population, 2,500.)—The city of Superior, in Douglas County, Wis., is the present eastern terminus of the Wisconsin Division of the railroad, situated on the Bay of Superior, near the extreme western end of the great lake. It shares, with the neighboring city of Duluth, the advantage of being at the head of navigation of the great lake system of the United States. Its sheltered, deep and spacious harbor, coupled with the fact that a rich and extensive region is tributary to its commerce, will doubtless make Superior an emporium of trade and industry. Merchandise from Europe may be transported to the wharves without the necessity of breaking bulk in transit, and from this natural inlet to the great and fertile regions of the Northwest goods may be distributed with a minimum of cost and delay. The entrance to Superior Bay, 500 feet wide, is flanked on either side by substantial piers and breakwaters, that have been constructed by the Government of the United States. Beyond the entrance the bay expands to a width of more than a mile, and to a length of nearly ten miles. This fine harbor furnishes twenty-five miles of available coast line for the accommodation of commercial and manufacturing establishments, every point of which is well protected from the most severe winds that vex the outer sea. The shores of this bay are already marked, at many places, with extensive flouring and lumber mills and other factories, and there is a constant growth of industrial enterprise. The city of Superior, fronting upon the bay, was projected and laid out before the civil war by a company of well-known citizens, among whom may be named: Wm. V. Corcoran, of Washington, D. C.; Robt. J. Walker, of New York; Geo. W. Cass, of Pittsburgh, Pa.; Horace S. Walbridge, of Toledo, O.; D. A. J. Baker, of St. Paul, Minn.; James Stinson, of Chicago, Ill.; Senator Bright, of Ohio; Senator Beck, of Kentucky, and J. C. Breckenridge, of Kentucky, formerly Vice-President of the United States. Stephen A. Douglas, after whom the county is named, was also interested in Superior in its early days. The idea, therefore, of building a city at the head of Lake Superior, on the bay, is not altogether new, although the development of the plan was left, on account of intervening national disturbances and financial complications, to the later generation which is now engaged in maturing it. Superior is already in communication with both eastern and western points by rail. The Northern Pacific Railroad here connects with the Chicago, St. Paul, Minneapolis and Omaha system, for Chicago and the East, as well as with several lines of steamers running to and from United States and Canadian ports on the lower lakes. As an evidence of the present prosperity of the city it may be said that over $700,000 were expended in improvements during the season of 1882. The Northern Pacific Railroad Company has built a magnificent dock for the convenience of its large freight business, and has made the plans and specifications for two similar docks, which are to be constructed in the immediate vicinity, in order to meet the wants of a rapidly growing commerce. An elevator of great storage capacity, to cost $250,000, is in course of erection by a company, organized under the laws of Minnesota, for the purpose of receiving a portion of the output of grain from the wheat-growing regions on the line of the Northern Pacific Railroad. Another company has been incorporated, under the laws of Pennsylvania, with the object of handling coal and iron at this point, where, also, a lively business in salt, lime, cement and other bulky lake freights is carried on.

Among its manufacturing and industrial establishments Superior has two large lumber mills with a capacity of 100,000,000 feet of planed lumber a year; a shingle mill able to produce 1,800,000 shingles a week; several grist mills and machine shops, and two brick yards, while many other enterprises of a similar character are projected. The town has eight churches;

five school buildings, valued at $30,000; a court-house, and other substantial county buildings; six hotels, two of which are first-class; a national bank; a telephone system and two enterprising newspapers. The various trades, occupations and professions are, also, well represented.

The Northern Pacific Railroad enters the city on the south, by Newton Avenue, and runs down to the bay, whence one branch is laid along the harbor line to Conner's Point and St. Louis Bay, and another across the Nemadji and Alloues rivers to Itasca Street switch, on the east side of the city, tapping all the mills and manufactories in its course.

Among the principal scenic attractions that are visible from the cars between Superior and Northern Pacific Junction, on the main line, are the long trestle bridge; Silver Creek, a famous trouting stream; Rock Cut, in the Copper Range; the two Pokegama Rivers, which are crossed by lofty bridges, and where the views are wild and grand; the Nemadji River, quiet and gentle, as it winds its way through pleasant woods and fertile meadows; Alloues or Bluff River, with its precipitous banks and deep and rapid stream; the Bay of Superior, land-locked by Minnesota and Conner's Points; St. Louis Bay and the Bay of Shelter, at the mouth of the St. Louis River, flanking the city on the right; and, beyond all, the sublime spectacle that meets the eye at the Dalles of the St. Louis River, where the stream tumbles in impetuous fury from the craggy range above. There are, also, minor objects of interest in the neighborhood of Superior. These are the Bay of Alloues, on the left of the town, a place noted for its quiet waters and the fine hunting about its shores; the rice fields of the St. Louis and Nemadji rivers; the picnic woods of Wisconsin Point; the commanding view from the Edwards Copper Mine of Lake Superior and its north and south shores; the fish-breeding ponds of the Aminicon River; the beautiful Brule River glens; the white birch plantations on the banks of Poplar and White rivers, and the Aminicon, Nimekagon and Nebagamain lakes. One of the greatest of natural scenes is the magnificent Black River Fall, ten miles south of Superior, and only a few miles distant from the track of the Northern Pacific Railroad. This cataract is 210 feet in height, and the water is of ebony blackness. It is one of the most interesting and awe-inspiring sights in the Northwest. The Aminicon Falls, 110 feet high, are also noted for their grandeur. The forests in the vicinity abound in deer, bear, and other large and small game, and the streams and lakes afford excellent trout fishing. The salubrious air, the beautiful scenery, and the fine opportunities for sporting and out-door life, combine to render Superior a pleasant summer resort, which visitors from the South and East frequent in large numbers.

A little village with a good hotel has grown up on Nebagamain Lake, and another summer hotel is to be built on one of the points of the bay. There are some relics in the vicinity of the old posts of the Hudson Bay Company, which tourists of an exploring turn like to visit. *Walbridge* and *Carlton* are yet unimportant stations.

Thompson—(twenty-two miles from Superior; population, 300)—situated near the head of the Dalles of the St. Louis River, is a favorite resort for tourists and pleasure seekers. It is the county-seat of Carlton County, and has a public hall, school, hotel, church, two stores, saw mill and shingle mill—its chief interest being the manufacturing of lumber. The scenery along the river is of a most varied and picturesque character.

A CHIPPEWAY CHIEF.

Benjamin Louis Coker

Josephine D. Roy
Quebec

Daniel Clark M.D.
Asylum
Toronto
Ont.

G.L. Clark
Toronto

Duluth, Minnesota.

Minnesota Division.

FROM DULUTH TO BRAINERD.—DISTANCE, 114 MILES.

Duluth.—(Population, 14,000.)—This is an important terminal point of the railroad, and is the most western of all the cities which lie on the great chain of North American lakes. It is the county-seat of St. Louis County, and one of the most flourishing and rapidly developing places of the Northwest. The financial storm of 1873 swept over Duluth, and sadly checked its growth during many years; but it has fully recovered from the shock, and is now advancing with rapid strides to a position of commercial and industrial preëminence. Its principal trade is in the famous hard wheat, the market for which product is constantly enlarging. To handle the grain there are three elevators at Duluth, with a united capacity for storing 2,660,000 bushels; and, during 1882, not less than 4,191,133 bushels were received and shipped. The harbor is deep, sheltered, and capacious enough to accommodate a large number of vessels, and the wharf facilities are excellent. Maritime business is fast assuming great importance, the arrivals and clearances during the navigation season of 1882 having reached 1,691, or an average of four per day, and the transportation of freight by vessels on the lake footed up an aggregate of 700,000 tons. The total receipts and shipments by rail during the period named amounted to 853,269 tons. Duluth is also the centre of a large lumber industry. In 1882 the twelve saw mills of the city manufactured 84,218,153 feet of lumber, 21,363,000 shingles and 10,295,000 laths. Since that date the capacity of the mills has been so increased that 110,000,000 feet of lumber, and a correspondingly larger quantity of shingle and laths, may be turned out. In 1882 no less than 503 new buildings were constructed, at an aggregate cost of $1,489,000. Duluth has excellent schools, many fine churches, imposing business houses, elegant residences, a well conducted daily newspaper, and other features of a thriving city.

The historic period of Duluth dates from 1856-7, in which years J. B. Culver, Wm. Nettleton, Luke Marvin, Sidney Luce, Geo. R. Stuntz and others procured letters patent to the lands on the Minnesota side of the bay, and founded a village on the site of the present city. But the first real impetus given to the place was in 1869, when the St. Paul and Duluth Railroad, then known as the Lake Superior and Mississippi Railroad, was nearly finished, with Duluth as the lake terminus. About this time the erection of the first substantial business blocks in the place was begun. Meanwhile, General Geo. B. Sargent had established himself at Duluth, as the agent of Jay Cooke & Co., and as the advance guard of the Northern Pacific Railroad. When, in 1870, this Company began its great work of building a railroad across the continent, with Duluth as a base of operations, it caused the town to leap at once into prominence as a depot of supplies, a manufacturing point and wheat market. With the completion of the railroad to the Red River of the North, Duluth became an important port of entry for the Canadian transit trade to Manitoba and the Saskatchewan Valley. From 1870 to 1873, Duluth was the scene of great commercial and some manufacturing activity. During this period the foundation of iron industries on a large scale was laid; a blast furnace was built, and foundries and car shops established—car wheels having been cast in the foundries for the freight cars that were built in the shops; while boiler furnaces, and other heavy kinds of iron work, including stoves, were successfully manufactured. But the great commercial crisis of 1873-74 intervened, with its consequent cessation of all railroad operations, and ruinous deterioration of the iron interests generally, putting an effectual end to all manufacturing enterprises, and not only leaving the town commercially flat, but burdened also with a large bonded indebtedness. This debt was incurred to a great degree by the cutting of a canal through Minnesota Point into the harbor, and by the building of a useless but expensive dike across the bay to evade an injunction suit brought by the State of Wisconsin against the cutting of the canal in question. Under this vast burden of debt, Duluth staggered through five years of great commercial depression. Meanwhile, the Northwest, under the impulse of the revived energy of the Northern Pacific Railroad, began to develop into a great grain field, the country at large gradually recovered from its financial shock, and capital again sought investment. At Duluth large grain elevators were built; harbor slips were cut to accommodate the constantly augmenting merchant marine list, coal docks and freight houses were constructed, and lumbermen, becoming interested in the country, began the erection of first-class saw mills. Duluth was now once more the scene of intense commercial activity, increasing from a town of 2,000 inhabitants in 1857-8, to a population of 14,000 in 1883.

If the above glance at the history of Duluth is interesting to the man of affairs, the situation of the town renders it no less interesting to the sportsman, on account of the unrivaled trout streams; to the invalid, for the bracing and salubrious climate; and to the tourist, for its grand and picturesque scenery.

Situated at the extreme head of Lake Superior, and covering the sides of a prominent elevation, Duluth presents a bold and picturesque appearance, as it is approached either by water or by rail. The city will probably centralize midway between Minnesota and Rice's Point, in the neighborhood of its most rugged features, and, as it grows, will continue to develop those picturesque effects which are the delight of the artist. Minnesota Point alone, with its scythe-like curvature, and splendid sweep, is a feature in the landscape worth seeing, and a good carriageway over the hills, from Duluth to Rice's Point, offers a place of vantage from which to behold a scene that for beauty, distance and effect is, perhaps, without an equal in the Northwest. From this drive, not only is a sweeping view of the lake, along the north and south shores, for some twenty-five miles to the eastward, to be obtained, but the gaze may also wander past the Bay of St. Louis, and over the marvelous labyrinth of crooks, windings, lakes and inlets of the Spirit Lake region as far as Fond du Lac to the westward. Spread out beneath the feet are the Bay of Duluth and Superior, between Minnesota and Rice's Point, with Superior in the distance, accentuated by the massive outlines of the Northern Pacific docks and elevators, the whole affording a variety of scenery which it would be hard to find elsewhere in a single view.

This part of Lake Superior is one of the most interesting points to the geologist on the North American Continent. It is on good grounds considered the oldest region in the world. The theory is that the formation of the lake is due to some great volcanic action, long prior to the ice period; perhaps that the lake itself was the mouth of a great volcano. Duluth is built on the rim of this lake basin, upon foundations of trap and conglomerates of every conceivable description, with seams of quartz and veins of iron, copper and silver often cropping out at the surface. The ancient lake bed extends some twenty miles above Duluth, over Grassy Point, Spirit Lake, and the bed of the St. Louis River, as far as Fond du Lac, around which the lake rim curves, enclosing a region of striking beauty. The chain of hills is here cut through by the St. Louis River, causing that wonderful series of rapids which, in a distance of twelve miles, have a fall of 500 feet through masses of slate, trap, granite and sandstone, and are fast becoming celebrated as the picturesque regions of the Dalles of the St. Louis.

The mean temperature of Duluth, during the summer, is as follows: June, 57° 9´; July, 61° 9´; August, 63° 6´; September, 58° 5. Summer visitors find here every convenience for fishing, hunting and sailing parties. Tourists and scientists usually have an abundance of time at their disposal, and are able at leisure to find out the most desirable localities. But there are many who come by lake, and have only a day to spare, or the brief period that a boat is waiting. To the latter class a trip to the Dalles of St. Louis, via the N. P. R. R., is one of the most profitable ways of spending the time. For the benefit of those who come by rail, and who delight in the "gentle pastime," a list of the trout streams on the north and south shores of Lake Superior, and their distances from Duluth, is appended:

NORTH SHORE.

NAME OF STREAM.	DISTANCE FROM DULUTH.
Lester River	5 miles.
French River	14 "
Sucker River	16 "
Knife River	21 "
Stewart River	32 "
Silver Creek	33 "
Gooseberry River	38 "
Encampment River	41 "
Split Rock River	48 "
Cross River	85 "

The Gooseberry River is considered the best trout stream on the north shore, then Split Rock, and Stewart and Knife Rivers, in the order named. Among the fine bays and islands most popular with tourists are Knife Island and Stony Point, Agate, Burlington and Flood Bays. Agate Bay, especially, is visited, and the name is very appropriate. Its shores are lined with agates, among an endless variety of other variegated and curiously colored conglomerates, all specimen chips from the neighboring rocks and hills, but worn more or less smooth by the perpetual friction and grinding of the wave-washed beach. The north shore is very precipitous, and abounds in fine scenery. Cascades and rapids are to be found on nearly all the above-named streams.

SOUTH SHORE.

NAME OF STREAM.	DISTANCE FROM DULUTH
Aminicon River	12 miles.
Bardon's Creek	20 "
Brule River	21 "
Iron River	26 "
Flag River	34 "
Cranberry River	40 "
Sand River	50 "

The scenery on the south shore is not so grand as on the north, and it is necessary to go further back from the lake to get trout, but Sand, Brule and Iron rivers are excellent fishing streams.

On St. Louis, and other bays in the neighborhood, are good trolling grounds for bass, pike, pickerel, etc.; and for the more venturesome, there is trolling for lake trout in the vicinity of the bays and islands on the north and south shores. Ducks and wild geese abound in the season on St. Louis Bay and river, while deer have been found in greater number the last two winters than ever before. On the south shore are enough wolves to make deer hunting very exciting; bear are occasionally shot both on the north and south shores. Passing the unimportant station of *Rice's Point*, at a distance of 4 miles from Duluth, we come to

Oneota—(Population, 200.)—This is an Indian name, meaning Beautiful Mountain. The town has a church (Roman Catholic), one store and a saw mill of 8,000,000 feet capacity. Brook trout are caught in small streams in the immediate vicinity, and pickerel in the bay adjoining the town.

Spirit Lake—(8 miles from Duluth.)—Here the St. Louis River widens and is called a lake, out of which rises an island which tradition says is haunted. The legend is, that in the early settlement of the Northwest, a captive white woman was cruelly tortured to death by the Indians, and her spirit was ever afterwards seen hovering around the place, threatening death and destruction to the red man. However this may be, the fact remains, that none of the Indians in the vicinity can be induced to put foot on the island.

The scenery westward from this place to Thompson, a distance of fourteen miles, is very grand. As the train pursues its way, now speeding through deep cuts, over yawning chasms and bridges reared to a giddy height, the foaming waters rush madly on, with a deafening roar, through the Dalles of the St. Louis far below, leaping from rock to rock, and enveloping them in clouds of spray. The river, in this stretch of about fourteen miles, has a fall of over 500 feet, and the views of the rapids and cascades from the car windows are exciting as well as picturesque in the extreme.

Fond du Lac—(14 miles from Duluth; population, 200.)—The town is beautifully situated on the St. Louis River, on the line of the St. Paul and Duluth and Northern Pacific Railroad. As a summer resort it has the attractions of fine scenery, and good hunting and fishing. It is one of the oldest settled towns in the State, the Hudson Bay Company having established a trading post here in 1790. Although now an unimportant station, it bids fair, at no distant day, to become a town of considerable business activity. Wood, ties, cedar posts, telegraph poles and brown sandstone, of which latter there is an inexhaustible supply, are the principal articles of shipment. The St. Louis River Water Power Co.'s boom, in which the logs are assorted for the Duluth market, is located here, and a fine chance offers to see the singular expertness which lumbermen attain in walking over the floating logs. The ease and rapidity displayed in skipping about over acres of footing so precarious are always a marvelous exhibition of skill, even to persons who daily witness the feat of stepping over revolving logs. The Water Power Company intend soon to construct a canal on the south bank of the river, for the purpose of furnishing power to mills to be erected, the sites being some of the most eligible in the Northwest. Fond du Lac has the ordinary supply stores, a shingle mill, a school-house and a post-office. There is a mineral spring here, which is said to be unsurpassed for its medicinal properties. The surrounding country is rough and broken, and covered with a dense growth of timber. Deer, moose, bear, wolf, lynx and fox are among the game animals to be found in the vicinity. Two miles west of the station are the celebrated fishing grounds of the St. Louis River. Here, each spring, upon the breaking up of the ice, thousands of pounds of fish are caught and shipped to Eastern markets. Close to the fishing grounds, in a sharp bend of the river, is the scene of the last famous battle between the Chippeway and Sioux tribes of Indians, in which the latter were completely defeated and routed with great slaughter, and their power forever broken in this region.

Northern Pacific Junction—(23 miles from Duluth, 131 miles from St. Paul, and 91 miles from Brainerd; population, 600.)—This is the junction of the St. Paul and Duluth and Northern Pacific Railroad, and a branch of the St. Paul and Duluth Railroad (known as the Knife Falls Branch), which runs six miles north to Knife Falls and Cloquet, where three saw mills are established, and large quantities of lumber are manufactured. Northern Pacific Junction has five hotels, one public hall, a church, good schools, and the county jail. It does a large business in supplying the numerous lumbering camps which are situated in the vicinity. Two saw-mills are here, which run summer and winter, and are supplied with logs during the winter by a logging train. The climate is dry, equable and healthy, the place being situated among the pines, where high winds do not reach it, although, in summer, the lake breezes cause an even, pleasant temperature. *Pine Grove*, 28 miles from Duluth, is only a side track. The same may be said of *Norman*, 33, and also of *Corona*, 39 miles from Duluth.

Cromwell—(45 miles from Duluth; population, 75)—is situated on a beautiful lake, which is stocked with pike, pickerel and perch. Cromwell has a section house, telegraph office, hotel and water tank. The principal shipments are wood and ties. Game: deer, bear, rabbits and grouse. *Tamarack*, 57 miles, *M'Gregor*, 66 miles, and *Kimberley*, 75 miles from Duluth, are small places of little importance except as points for the shipment of wood, ties, fence posts and telegraph poles, which are cut from the neighboring forests.

Aiken—(87 miles west of Duluth, and 27 miles east of Brainerd; population, 600.)—The town has three hotels, two of them being the largest between Duluth and Moorhead, three general stores, the usual shops, a public hall, churches, schools, two saw mills, a planing mill, a flour mill, a bank with a capital of $100,000, and a weekly stage line to Grand Rapids, Minn. A few miles north the Mississippi River has its source in *Itasca Lake*, in the vicinity of which an immense lumber trade is carried on, the trees being cut into logs and floated down the Mississippi to the Minneapolis mills. The Mud River rises twenty-five miles southwest, and flows through sixteen large lakes, which are full of fish; *Red Cedar Lake*, with its fifty miles of shore, and five other lakes of good size, situated four miles west of Aiken, are excellent places for hunting and fishing. *Crystal Lake* is distant two and a half miles south. *Lake Mille Lac*, twelve miles in the same direction, is noted for its beauty, and all are well worth a visit.

The country around the lakes is surpassed by none in point of attractiveness to the eye, being undulating and park-like. The glades and meadows are spangled with wild flowers in great variety, and the pebbly shores of the lakes, and azure, transparent waters, present a scene which impresses the beholder by its rare beauty. The hunting here is excellent. Elk may be found within seventy-five to one hundred miles north of this point, and in the immediate vicinity of Aiken are deer, bear, geese, ducks, pheasants, grouse and woodcock.

Visitors to this portion of Minnesota, desiring to see the red man in his wild way of living, may have their wishes gratified by driving out to the great and beautiful Mille Lac Lake and Chippeway Indian Reservation, about twelve miles from Aiken. The tourist who cares to penetrate still further up the Mississippi River, can do so by steam-boat, leaving this point once a week. *Cedar Lake*, 92 miles west of Duluth, is only a side track.

Turning the Tables.—Bishop Whipple says: "During the first survey of the Northern Pacific Railway, a party of men were lost, and at last came out half starved to the Indian village of Mille Lac. They went to Shah-bah-skong's house, and asked for food. The chief's wife prepared, at his request, an ample meal, and he then called his wife and children, and sat down to eat, leaving the white men as wallflowers to look on. After the meal was over, another table was spread, and the white men were invited to eat. They could do nothing but pocket the insult. After they had eaten, the chief said, 'My friends, when I was in Washington, the great father told me, that, if I wanted to be happy in this world, and to go to a good place when I died, I must keep my eyes wide open, and see how the white man did, and copy his example. I noticed that a rich white man never had a poor man sit at his table, and people of another color always waited. You are poor men to-day, of another color. I am the rich man now. I want to be happy in this world, and would not like to lose my chance to go to the good place, and so I asked you to eat at the second table.'"

Deerwood—(97 miles west of Duluth; population, 30)—is a favorite retreat for the hunter, and one of the wildest, least known and most beautiful points on the Northern Pacific Railroad. An unbounded forest stretches in every direction, in which deer and bear tempt the adventurous sportsman to share with the Indians the excitement of the hunt. The small streams and clear lakes, of unknown depth, invite the lover of the rod to make his camp here. The invalid, who craves repose, yet does not care to be too far away from the post-office or telegrams, finds here his Mecca. A small hotel has been built, and accommodation may also be found among the farmers at this point; or, if camping out is preferred, it is easy to obtain milk, eggs, ice, fresh vegetables and berries from the same source. Game and fish may be had for the taking. With such fare, under Minnesota skies, with his hammock swinging over beds of sweet fern, or his boat floating among beds of water lilies, who would not be happy?

In a radius of three miles, there are over twenty known lakes, whose waters fairly teem with muskelonge, pike, black bass, white fish, pickerel, croppies, wall-eyed pike, sunfish, rock bass, catfish, bull-heads and suckers. It is not uncommon to take pike weighing upwards of twenty pounds, and black bass six pounds, with a trolling spoon, while at the mouths of streams bass weighing from half a pound to two and a half pounds can be caught with the fly. The lakes vary in size from little gems, a few hundred feet across, to larger ones of several miles in diameter, many containing islands. Some of them have high rocky shores, pebbly beaches, and deep blue water; others, fringed with a growth of wild rice, are the feeding and hatching grounds of numbers of wild fowl. The more distant lakes can be reached by pony and buckboard, or by birch canoes, the latter carried over portages.

There is a little trading post at this point, which, from its various shipments of furs, fish, venison, game, maple sugar, cranberries, raspberries and huckleberries, gives a very good idea of the resources of the adjacent country. Here also the civilized Indians can be seen at their several occupations, from making maple sugar and birch bark canoes in the spring, to gathering wild rice in the fall, and hunting and trapping in the winter. The sportsman finds here in their season deer, with an occasional caribou, black and brown bear, wolves, foxes, coon, beaver, black and gray squirrels, the great northern hare, Canada grouse, wood ducks, teal, mallards and blue bills. Deerwood will probably be one of the favorite summer resorts of the many in northern Minnesota. *Jonesville* (108 miles from Duluth) is a side track.

Six miles beyond Jonesville the train reaches Brainerd, and unites at that place with the line from St. Paul.

[*A description of Brainerd will be found under the heading "St. Paul Division," page 45.*]

Gull River—(143 miles from St. Paul; population, 600.)—Gull River, so called from the river which runs through the town, is a lumbering point from which great quantities of lumber are shipped for building purposes. One of the largest saw mills in the State is situated here; also a sash and door factory, and a planing mill. There are two hotels, a general store, a school-house, and the necessary shops. **Gull Lake** lies four miles north of the town. This is another of Minnesota's beautiful lakes, abounding with fish of all kinds. There is a steam-boat on its waters which carries the tourist from eighty to one hundred miles around its shores. Two miles west of Gull River is

Sylvan Lake, also a very pleasant resort in summer. There are a great many deer, and some moose, in the neighborhood of these lakes. A moose was recently killed that weighed, when dressed, 800 pounds. Wolves and bears are also to be found. In the spring and autumn the rivers and lakes are alive with ducks and other water fowl. Years ago, one of the greatest battles between the Chippeways and Sioux Indians was fought here. "Hole-in-the-day," one of the Chippeway chiefs, was shot in this vicinity. "Bad Boy," so called by the Indians, because he saved many of the white settlers' lives, at the time of the Indian massacre, in 1862, lives here.

Pillager and Bath, respectively 148 and 154 miles distant from St. Paul, are side tracks, and passenger trains do not stop at these stations.

Motley—(158 miles west of St. Paul; population, 400.)—Motley has two hotels, a school-house, a church and a newspaper. It is situated in a lumbering district, and its two saw mills cut 12,000,000 feet of lumber yearly. Very little farming has yet been done in the neighborhood, but a few persons have settled here with the intention of cultivating the ground, as they find the soil fully as good as prairie land, and in a short time there is likely to be a prosperous farming community around Motley. There are several lakes near the railroad, and among them *Lake Shamiveau*, about six miles south of the town, and *Alexander Lake*, twelve miles distant, in the same direction, both affording very good fishing. This is also one of the best of hunting grounds, over 1,000 deer having been lately killed, together with a number of black bears. The few Indians remaining in this neighborhood are industrious; a large number of them, having given up their wild mode of life, are at work in the saw mills. The Indians are said to work faithfully, and accomplish fully as much as white men. Stages run Wednesdays and Saturdays to *Long Prairie*, eighteen miles south.

Staples Mills—(165 miles from St. Paul; population, 150.)—This place contains two saw mills and a grain elevator. The inhabitants are engaged in lumbering, cutting wood, railroad ties, piles, etc. Game is plentiful in the neighborhood.

Aldrich—(172 miles from St. Paul; population, 125.)—The town, situated on the Partridge River, a beautiful little stream with well defined banks and rapid current, thirty-six miles west of Brainerd, lies in Wadena County. The land surrounding the village is mostly covered with timber, consisting of birch, oak, maple, tamarack, spruce and pine. Large quantities of wood, railroad ties and piling are shipped from Aldrich, to supply the demands of settlers further west, and millions of feet of pine logs are floated down the river in the spring of the year. A high ridge of land runs north of the village, and upon it is the principal road leading to the outlying farms. The soil, a rich clay loam, produces large crops of wheat, oats, corn and potatoes, as well as garden vegetables. The farmers, an industrious, thrifty class of people, have the prairie openings well developed into good farms, and ship heavy crops of wheat, paying much attention also to stock raising and dairying. Aldrich is supplied with good stores, a well kept hotel, a saw mill capable of cutting 20,000 to 30,000 feet of lumber daily, an elevator and a side track large enough for ninety cars. This is a good point for game of all kinds, while the Partridge River and the neighboring lakes are well stocked with fish.

Verndale—(153 miles west of Duluth, 175 miles northwest of St. Paul, and 97 miles east of Fargo; population, 600.)—This town is pleasantly situated in Wadena County, in the Wing River Valley (one of the most fertile and beautiful valleys of the Northwest), of which it is the commercial centre. This valley is twenty miles in length, by five or six in breadth, and consists of a number of small prairies or openings, so admirably arranged by nature that almost every settler has timber and prairie. The village is about one mile east of the river in a beautiful opening, or small prairie, sheltered on the north and west by a dense grove of pines, while about two miles south and east can be seen the dark line of the Big Woods, which stretch away for many miles. Verndale is in Aldrich Township, and nearly centrally situated in the county. In its vicinity are many fine farms, the richness of the soil and the thrift of the inhabitants leaving nothing to be desired. Wing River, about one and a quarter miles distant, furnishes a fine water power, which is used in supplying the mills. There are four hotels, two banks, a newspaper, two public halls, a church, good schools, several general stores, as well as a flouring mill and a saw mill. A number of business buildings and residences were erected during 1882, at a cost of $41,525. A large portion of the people of the northern part of Todd County find Verndale their most convenient trading post. The best timber lands in

Wadena County, beside the Big Woods to the south, are adjacent to the village, and lumber is easily shipped from this point. The water power is also a great advantage, being the only one on the line of the railroad for a distance of twenty-five miles. These, and several other natural advantages, tend to make Verndale the commercial centre of a large tract of country. The products are wheat, oats, potatoes and garden vegetables. The shipments are flour, wheat and potatoes. Deer, bear, rabbits, grouse and pheasants invite sportsmen to the place, and the lakes abound in many varieties of fish.

Wadena—(183 miles west of St. Paul; population, 1,100.)—This town is the county-seat of Wadena County. It is a lively, progressive place, incorporated and governed by a village council composed of five members. The streets are wide and well graded, while the board walks are cut on a liberal pattern, extending from the door yards to the street. The greater part of the town lies on the south side of the railroad. The first line of buildings, which are mostly devoted to business purposes, are about two hundred feet from the track, leaving an open and unoccupied space extending along the entire front of the town site. This ornamental, but too often overlooked feature, adds very much to the attractiveness of the town, at the same time answering the purpose of a public space for the enjoyment of base ball, lawn tennis, and other outdoor sports. The houses are of a good class, the building material, of course, being mostly wood, but a brick yard has recently been established near the town, which supplies a good quality of red brick. The citizens are mostly of American birth from the older States. Water is excellent and easily obtained. The educational facilities are good, there being a common and a graded school. Churches are well represented by five religious denominations—the Congregational, Episcopal, Methodist, German Methodist and Lutheran. The Congregationalists and Methodists have neat and commodious church buildings, which are well attended, and the Episcopalians are about to erect a fine church edifice. Peake's Opera House affords a convenient place for dramatic and musical entertainments, lectures, church fairs and dancing parties. The country adjacent to the town is a slightly rolling prairie, dotted at intervals with picturesque groves and strips of timber. Oak, poplar, birch and ash are the most common growths. A few miles north of the town begins the timber line, beyond which lie some of the famous logging camps of Minnesota, where are found large tracts of white and yellow pine. Wadena is therefore a convenient shipping point in winter for cord wood, ties and piling, in fulfillment of large contracts made with the railroad and individuals. This business is also very beneficial to the farmers, giving employment to themselves and their teams when the usual farm work is at a standstill. Wadena depends not alone for its support on the county wherein it is located; but, being favorably situated, draws a great amount of trade and business from Todd, Otter Tail, Becker and Cass Counties, which are immediately adjoining. The fact that the town is so important a shipping point encourages business enterprises, among which may be mentioned two banks, the Merchants' and the Bank of Wadena; four general stores; the Clayton Manufacturing Company, devoted to the production of plows and general foundry work; a post-office; three drug stores; three hotels, besides several boarding houses, one patent roller flouring mill with a capacity of 100 barrels per day; two hardware stores, two grain elevators, three agencies for the supply of reapers, binders and other agricultural implements, together with stores of a miscellaneous character. The total amount of building improvements in the village of Wadena during the year 1881 was $35,000, and in 1882, $80,000. The products are wheat, barley, corn, oats and potatoes. A semi-weekly line of stages runs to the agricultural village of *Wrightstown*, twelve miles distant, with 175 people, and to *Parker's Prairie*, twenty-five miles distant, with 350 inhabitants.

Wadena is also a coal and water station for locomotives, as well as the northern terminus of the Northern Pacific, Fergus and Black Hills Railroad, which is now running out into central Dakota, and is fast extending its line further westward. This road also furnishes the citizens of Wadena with convenient and easy access to Battle and Clitheral Lakes, thirty miles southwest, which are popular picnic and fishing grounds.

Wabasha's Logic.—On one occasion the Dakota Indian chief, Wabasha, was rebuked by a missionary for having engaged in a scalp-dance after he had returned from a raid upon his enemies, which resulted in the capture of several ponies and a single scalp. The Indian heard the rebuke patiently, but was not convinced that he had done wrong. He said: "White man goes to war with his own brother, kills more men than I can count on my fingers. Great Spirit looks down from sky and says: 'Good white man; got my book; me love him very much and have good place for him by and by.' Dakota got no Great Spirit Book; go to war; kill one man; come home; have scalp-dance; Great Spirit mad. Wabasha does not believe it." The red man's logic was unanswerable.

Northern Pacific, Fergus and Black Hills Branch.

FROM WADENA TO WYNDMERE.—DISTANCE, 104 MILES.

This branch railroad is to be extended to the precious mineral regions in the Black Hills of southwestern Dakota, opening up a fine agricultural region many hundred miles in area on its route. At the time of writing the track was being rapidly laid westward.

Deer Creek—(10 miles from Wadena; population, 150.)—This place, which was founded not more than a year since, is situated in the midst of a good wheat-growing and timber country, and promises to develop into a thriving town. It has already two stores, one blacksmith shop, an elevator and an hotel. Plenty of game is found in the vicinity, such as deer, rabbits and partridges. *Parkton* is a new station, four miles west of Deer Creek.

Henning—(18 miles from Wadena; population, 300).—The town contains about fifty buildings, of which three are stores, carrying on a general business, and two hotels, the American House and Scandinavian Hotel, both offering very good accommodations. There are also two blacksmith shops, three carpenter shops, a meat market, a hardware store, etc. Here the Mississippi and the Red River of the North almost interlock. Forty rods east of the village site runs Leaf River, which empties into the Mississippi; and the same distance west the streams flow into the Red River of the North. Two miles south of the village are the Leaf Mountains or Painted Hills, rising about 200 feet above the plains, making an elevation of about 1,700 feet above the level of the ocean. From these eminences a beautiful view is presented of the surrounding country. Henning occupies a central location to three of the finest lakes in the park-like region, *viz.*, *Inman Lake*, on the east, with its crystal waters and heavily wooded shores; *East Battle Lake*, on the west, with its islands, bays, rocks and headlands, embowered amid the shades of the primeval forest; and *Leaf Lake*, on the north, with its deep, clear waters, and its shore line of twenty-five or thirty miles bordered by thick woods. There are several other charming lakes, such as *Round Lake*, with its white gravelly beaches; *McDonald*, *Buchanan* and *Otter Tail Lakes*, the latter the largest of all, being ten miles long by three miles wide. Each of these lakes has its own distinctive attractions, making it the choice resort for either the invalid or the pleasure seeker. These lakes all abound in many kinds of excellent fish, such as white fish, pickerel, pike, catfish, and black and rock bass. This region has always been the resort and breeding ground of large numbers of water fowl, and no less than seventy varieties of birds have been found here.

Vining—(24 miles from Wadena.)—This station lies in the midst of a good grain-growing country, and the region is well timbered with oak and maple.

Clitherall—(29 miles from Wadena; population, 150.)—This new town, half way between Wadena and Fergus Falls, is situated near three of the finest and largest lakes in the renowned Minnesota park region— *Clitherall Lake*, and the two noted *Battle Lakes*, west and east, respectively. There are two good hotels, three stores, in which are kept a complete assortment of general merchandise; one drug store, one hardware store, two grocery stores, one boot and shoe store, a large elevator, a lumber yard and a blacksmith shop. Clitherall Lake, a great pleasure resort, is a beautiful sheet of water, somewhat in the shape of the capital letter Y, extending from northeast to southwest, about four miles in length, with an average depth of sixty feet. It teems with every species of fish known to the Western lakes, from the monstrous buffalo of forty and fifty pounds avoirdupois, or the shy pickerel of twenty-five pounds, down to the beautiful perch of a couple of ounces. The lake is also haunted by water fowl in great numbers, from the pelican and goose to every species of ducks. On its shores there is a small Mormon settlement, the oldest in Otter Tail County, the people having made their homes here as early as 1865. They are followers of Joseph Smith, but bitter denouncers of polygamy and their cousins at Salt Lake. Their settlement is one mile and a half from the station, and is finely situated in a beautiful grove of oaks on the north shore of the lake. They have about five hundred acres under cultivation, and the railroad runs through their fields in sight of the settlement.

South of Clitherall, for ten miles, stretches a grand prairie, and he must indeed be a poor shot who cannot here bag as many grouse as he wants. The Leaf Mountains are the favorite haunts of deer, which are killed by white hunters, in great numbers, every autumn. The Indians say that these mountains have been visited every year by them, in pursuit of deer, as far back as their oldest people can remember. Not even the presence of the white man and the railroad can drive the Indian from his "hunting ground." Even now, at all seasons of the year, the tourist can see here and there a wigwam on the north shore of the lake, and the eyes of a shy papoose peeping at him from behind a bush. The drives from Clitherall are very fine. Hotel accommodations are good for a limited number, and teams, with experienced drivers in charge, may be had at any hour by excursionists.

Battle Lake—(33 miles from Wadena; population, 600.)—Ere beautiful Lake Clitherall is lost to view as the train speeds along through pleasant groves and picturesque scenery, it rounds a high bluff, and another picturesque sheet of water is seen, covering an area of four by nine miles. This is the well-known Battle Lake. The town of Battle Lake lies at the west end of the lake, and a large amount of wheat is marketed here. There are two elevators, one hotel, a school-house costing $5,000, and a church, while a steam flouring mill is in course of construction. A look-out has been erected by the Northern Pacific Railroad, the view from which is magnificent. Seventeen beautiful lakes can be seen within a radius of five miles, all of which are well stocked with fish. Besides these there are many ponds where, during spring, summer and autumn, aquatic fowl are abundant. There are two Battle Lakes, *West Battle Lake* and *East Battle Lake*. West Battle Lake, the queen of Otter Tail County lakes, lies one mile north of the station, and is the largest of the three lakes named. It is a favorite resort for fishing parties, and the finny tribe seems inexhaustible. This lake has an average depth of seventy-five feet. A steamer, sail-boats and numbers of row-boats ply its laughing waters. *East Battle Lake*, the lover's retreat and the pleasure seeker's home, is hidden among the islands and woodland hills, and is renowned for its romantic scenery. The lake is quite irregular in form, its shores being broken by grottoes, dells, lovely little coves and bays. It is about four miles long and from half a mile to two miles wide, containing three large islands well suited to picnic parties. It is the natural home of wild ducks, which congregate here in the spring and autumn in countless numbers.

The Battle Lakes take their name from a famous and bloody conflict which was fought on the neck of land that divides their waters, between the Chippeway and Sioux Indians, in which the former won a dearly bought victory, killing every one of their enemies, but losing 500 of their own warriors. The battle ground is only a mile and a half from Clitherall, where the fortifications, breastworks, rifle-pits, and even the mounds over the graves, still remain as a record of the bloody and fatal strife between the savages for the possession of this most coveted hunting ground. On the north side of the lakes is still another earth fortification, where at some time another terrible battle was fought between the Indians. A breastwork, in circular form, encloses about an acre of ground, and inside the circle are a number of rifle-pits. Arrow-heads, shells and other relics have been found in this place.

Game is very abundant, and the sportsman has the advantage of not being troubled with mosquitoes, as the cool prairie air and lake breezes seem to drive them into the deeper woods, and good repose in the balmy air of a summer night may be enjoyed in a hammock stretched from tree to tree.

Maplewood—(39 miles from Wadena.)—This is a new place, but growing rapidly.

Underwood—(41 miles from Wadena; population, 75)—is an enterprising little town about ten miles east of Fergus Falls. It nestles amongst hills and beautiful lakes, which exhibit very fine scenery. The country adjacent cannot be surpassed for richness and productiveness of soil. The climate is splendid, and the summer season sufficiently long to mature all crops. The land not already taken is now open for settlement, and good fertile farms can be purchased at reasonable figures. The settlers have the advantage of an abundance of hard wood timber, and find lucrative employment in shipping wood to Western markets. The town contains one extensive chair factory in full operation, three stores, one elevator, with fair prospects for another, and a very fine depot. The inhabitants consist principally of Scandinavians, who have been settled here for some time. They are industrious and enterprising, and are in very comfortable circumstances. The whole country is dotted here and there by beautiful lakes, varying in area from two to twenty square miles. These lakes abound in varieties of fish, such as pickerel, pike, bass, etc., and other favorable resorts for the tourist. Large flocks of ducks and geese resort to them in spring and autumn, thus affording excellent shooting.

Fergus Falls—(52 miles from Wadena; population, 5,000).—The site of this city only ten years ago was a smiling wilderness; now it is the county-seat of Otter Tail County, with a busy and energetic community.

The city is three miles square, and is built up more or less for nearly two miles up and down the Red River, and over a mile in breadth north and south. To the north, overlooking and protecting the valley, are groves of timber, through which stretch narrow strips of prairie. South of the river the land is for the most part prairie, on which are several planted groves of rapidly growing trees. The growth of Fergus Falls has always been steady, but since the place has been opened to the world by rail the increase in population has been rapid. The principal street, Lincoln Avenue, is built up compactly on both sides for nearly half a mile, and business overflows thence up and down the cross streets. The business blocks are large and substantial, and the city is fast attaining a high commercial standing. Within an area of two miles north and south, by three miles east and west, are six distinct water powers, with over eighty feet fall. The Red River at this point leaves a high upland region and descends a distance of over 200 feet in a few miles to the level of the Red River plain, furnishing 10,000 horse-power, which is used for milling and manufacturing purposes. The favorable situation of Fergus Falls at the southern end of the celebrated Red River Valley, surrounded by a rich, well developed agricultural and stock-raising country, and in the midst of the famed Park region of Minnesota, gives the place a front rank among the thriving towns of the Golden Northwest. Good water power privileges are offered for sale or lease at reasonable figures, and manufacturing enterprises are being pushed with zeal. In close proximity to the town are numerous summer resorts, and the lakes, abounding in fish and game, yearly attract numbers of tourists and sportsmen. From Fergus Falls a branch line of the St. Paul, Minneapolis and Manitoba Railroad runs twenty-two miles past Elizabeth to Pelican Rapids, the latter a prosperous village of 600 inhabitants. *Ames* and *Everdell* are new stations on the Fergus Falls branch, which are gradually growing. The next important place is

Breckenridge—(77 miles from Wadena; population, 1,000.)—This is the county-seat of Wilkin County. It is situated on the western border, a little south of the middle of the county, on the Red (formerly Otter Tail) River, and on the east side of the Bois de Sioux. Wilkin County contains about twenty townships, all of which are excellent agricultural lands. The soil is a deep drift of dark color resting on a subsoil of clay. Intermixed with the surface soil are immense deposits of minute lacustrine shells, containing large proportions of phosphates, and its upper portion is composed largely of the *débris* and ashes of vegetable matter. In 1857 a cabin was built on the town site from timber cut on the north side of the river. Shortly afterwards the ground was surveyed, and a town plat, three miles long and two miles wide, was laid out. A contract was let for building a hotel, thirty by sixty feet, three stories high, and the building was finished in 1859. Machinery for a steam saw mill was hauled hither from St. Cloud in 1859, and a mill and several houses were erected; in fact, Breckenridge became thus early a prosperous place. But it received a check to its progress from the Indian outbreak in 1862, at which time it was burned and eight of the inhabitants slain. Mr. James C. Rice, who had charge of twenty-five wagons belonging to Mr. Burbank, of St. Paul, reached Breckenridge in 1862, on his way to Pembina, with supplies for the posts of the Hudson Bay Company. He succeeded in getting into the stockade near Breckenridge with his men and teams before the Indians attacked the post. There was a company of eighty soldiers in the fort, all raw recruits, and the battle was maintained for two days. As the Indians could make no impression on the works, they finally withdrew, after losing several of their number, but without inflicting much damage upon the whites.

It was not until some eight or ten years had passed that the rebuilding of Breckenridge was begun, nor did it again become a place of any importance until 1873, when the St. Paul and Pacific Railway was finished to that point, since which time it has had a steady and substantial growth. Breckenridge has a round-house and repair shop, two general stores, a court-house, a hardware store, a boot and shoe store, a drug store, two elevators, a warehouse, four hotels, two churches, a bank, a feed store, a machinery dealer and a newspaper. It has a fine school building, costing $10,000, built of brick with stone trimmings. The Minnesota House, the largest hotel in Breckenridge, is situated on the corner of Fourth Street, and has accommodations for a large number of guests. Several good and extensive farms are in the neighborhood of the town, yielding heavy crops of wheat and oats. Breckenridge has an excellent water power.

Wahpeton—(78 miles from Wadena; population, 2,000.)—This town, situated on the west side of Red or Otter Tail River, just below its confluence with the Bois de Sioux, is the county-seat of Richland County, one of the best agricultural counties in Dakota Territory. It is forty-six miles south of Fargo, and at the head of navigation on the Red River of the North. Wahpeton has a water power, formed by the Otter Tail, with a fall of sixteen feet, furnishing a steady and reliable volume of water. A company has been formed, composed of the leading citizens of Wahpeton, with a capital of $75,000, to begin its improvement. Thirteen years ago the first claim but was put up on what is now the town site. In 1873 a trading house was established, and traffic was carried on with the Indians, who occupied nearly the entire country from Big Stone Lake to the British Dominion for miles on both sides of the river. In 1876 the place was laid out in lots, and soon afterwards was recognized as an eligible town site. Wahpeton has now a fine opera house; a court-house, erected at a cost of $30,000; two good newspapers; several churches; two banks; school buildings, and five hotels; while all branches of business are well represented. There are an elevator of 100,000 bushels capacity, two large grain warehouses, a steam flouring mill, a steam factory and repair shop, two railroad depots and four lumber yards. The town is in the midst of an agricultural country of superior fertility, and ranks, as a commercial centre, among the first in eastern Dakota. Its growth has been rapid and substantial, and its further development is assured by the establishment of new industries. The County of Richland, with about 8,000 inhabitants, nearly all of whom have settled there within five years, and fully half since 1880, has an area of 1,332 square miles, or 850,000 acres, of which about 65,000 acres are under cultivation. The farms, with the exception of four or five "bonanza farms," are owned by settlers whose tracts consist of from 160 to 320 acres. There are various kinds of timber in Richland County, consisting of oak, ash, elm, box-elder, linden and cottonwood. West of Wahpeton several new stations have been established, the principal points being *Ellsworth* and *Wyndmere*.

Minnesota Division.

[*Continued from page 65.*]

Bluffton—(187 miles west of St. Paul; population, 600.)—This town, very near the divide between the Mississippi and Red River valleys, is situated on Bluff Creek, a branch of Leaf River, into which it empties about half a mile below the town. In the month of August, 1878, Bluffton was organized and given its name, a school district having been formed out of it and a part of Compton. The first school was held in a store, the goods on one side and the school on the other. The town at that time contained thirteen voters. It now has a hotel, a church, a public hall, a school, blacksmith and wagon shop, a saw mill, a planing mill, a grist mill, a flouring mill, an elevator and a post-office. Its principal industries are the raising of wheat and the shipping of wood, ties, lumber, wheat and flour. Small game and fish are plentiful, and deer is abundant in this region. *Amboy*—(190 miles from St. Paul)—is simply a side track.

New York Mills—(195 miles west of St. Paul; population, 500.)—The inhabitants of this town are chiefly of Russian and Norwegian nationalities, and mostly engaged in the manufacturing of lumber. There are a steam saw mill, with a capacity of 30,000 feet per day, one hotel, six general and other stores, a school and a church. Three miles northward is another steam saw mill, turning out 15,000 feet per day, and, ten miles beyond, is still another, with a capacity of 30,000 feet per day. The three mills, combined, give employment to over 200 men. The surrounding country is well timbered, and the soil is a rich black loam. The region is interspersed with small lakes, which are well stocked with fish. In the proper season deer are very plentiful, also small game of various kinds. The population of this town is steadily increasing. *Richland*—(5 miles westward)—is simply a side track.

Perham—(206 miles west of St. Paul; population, 1,000.)—This town, situated in the northeastern part of Otter Tail County, on an open prairie of four or five miles square, is one of the most prosperous places on the line of the Northern Pacific Railroad. Every branch of business is well represented, and manufacturing enterprises are flourishing. Among the latter are a carriage and wagon factory, a sash, door and blind factory, and a barrel and stave factory. The town supports four good hotels, prominent among which are the Merchants' and the Grand Pacific, the latter a new three-story brick building. There are five churches, several good schools and a newspaper. Perham's greatest pride is her five-story Steam Prairie Roller Mills, fifty-four by fifty-six feet in dimensions, and worth $60,000, with a capacity of 250 barrels of flour per day. In connection with the mill there is a large warehouse, thirty-six by forty-four feet in size, with a capacity of 35,000 bushels. The Northern Pacific Railroad Company has three large buildings, consisting of a newly erected baggage and express room, a large freight warehouse, and a neat and commodious depot, with waiting rooms and ticket office. The scenery about Perham is attractive. In coming from the east, for some distance nothing can be seen but pine forests, which suddenly open into a beautiful rolling prairie, through which the famous Red River of the North passes. To the right, only a short distance away, lie two beautiful lakes, called *Big* and *Little Pine Lakes*. The latter is about two miles wide and four miles long, while the former is nearly three times as large. The view from the passing train is very pleasing.

After leaving Perham there are lakes without number, which, to travelers from Eastern cities, would be considered marvels of beauty. All of these lie in sight of this thriving town. They are now becoming popular, and many tourists spend the summer on their banks. Among these resorts is *Otter Tail Lake*, four miles wide and eleven miles long. It is situated eight miles south of the town. *Marion Lake*, three miles distant, in the same direction, is perhaps three-quarters of a mile in diameter, and nearly circular in form. No better hunting ground can be found in the Northwest than that surrounding Perham. The lakes are full of fish of every description, including pickerel, pike, muskelonge, black and rock bass, catfish, sunfish and white fish. In spring and autumn ducks and geese are killed in great numbers. During the season the prairie and groves are alive with quail, grouse, swan, brant, woodcock, prairie chicken, partridge, snipe, curlew and rabbits. In early winter the deer, elk, moose and bear are an easy prey to the sportsman.

A Hunting Adventure.—A couple of Perham farmer lads, 16 to 18 years of age, were recently annoyed by some animal that helped itself at night from their corn crib. Becoming tired of such proceedings, and determined upon capturing the intruder, two old muskets were loaded for the occasion, two butcher knives sharpened to a fine edge, and fastened to their waists, and the boys started out. A light fall of snow the night previous made following the trail an easy task, and they kept on for three or four hours, until they were about ready to give up. At this juncture, directly in front of them, arose an immense moose. The boys, never having seen anything of the kind before, collapsed, the guns falling from their trembling hands. The moose walked off a few rods, then turned, and, with a disdainful look, made the woods ring with his bellow. The oldest boy cried out to his brother: "Why don't you shoot?" Thus aroused, the little fellow hastily picked up his old musket, and fired. After the smoke had cleared away the animal was seen struggling on the ground, and the boys soon dispatched him. The next day they came to town with their game, hauled by a yoke of cattle. The animal weighed 900 pounds and the captors received $65 for their prize. Hunters may learn from the above that game of no small dimensions is to be shot near Perham. Upon the lakes before mentioned are numerous yachts and row-boats, and in early spring steamers are placed on Big and Little Pine and Otter Tail Lakes. There are beautiful spots in the groves, where camp life can be enjoyed, and where the tired sportsman can recuperate his exhausted energies.

Two miles from the town is an Indian village. As a general thing, the Redskins are not afraid of work. During the winter the men take employment in the pineries, and get good wages, while the squaws look after their domestic affairs. They seem to have plenty of money, and always set the good example of paying cash for their goods.

Frazee—(217 miles west of St. Paul; population, 1,500.)—Frazee City, situated in Becker County, was established about eight years ago. It boasts of having the largest flouring mill west of Minneapolis, the product of which is shipped to all parts of the world. In addition to the flour mill there is a large saw mill, which is supplied with timber driven down the Otter Tail River from ten to twenty miles. There are two hotels, one public hall, a good school and a grist mill. Frazee City is surrounded by a first-class farming country, which is fast being populated by a thrifty and enterprising class of farmers, most of whom are in good circumstances. Otter Tail River, running through the town, is full of all kinds of fish, and so are the numerous lakes that find an outlet through this river. Opportunities for hunting bear, deer, wolf, fox, lynx, otter and beaver are exceptionally good; and also for shooting geese, ducks, grouse and pheasants. *McHugh*, five miles beyond, is only a side track.

Detroit—(227 miles west of St. Paul; population, 1,400.)—Detroit, the county-seat of Becker County, is situated in a beautiful timber opening, the surface of which is gently undulating, the soil being of a sandy nature. The village organization comprises the entire township, six miles square of territory, and its affairs are governed by a village council, consisting of a president, three trustees, a recorder, treasurer, two justices of the peace, a constable, a marshal and an assessor. The organization, under a village charter, was perfected in the winter of 1880, by act of the Legislature. Half a mile east of the village runs the Pelican River, which stream is the western boundary line of what is known as the "Big Woods" of Minnesota. To the west there is but little timber, and on the north the country is about equally divided between timber and prairie land.

South of Detroit lies what is known as the Pelican Lake country, one of the finest, as well as the most fertile and beautiful, sections of Minnesota. The surrounding region is very productive, and each year the farmers are blessed with abundant crops, for which a good and ready market is always found. The advantages of Detroit are many, and its prospects for the future are flattering. Its abundance of excellent oak, maple, elm, birch, basswood, tamarack and ash timber, suitable for the manufacture of all articles made from wood, invites industrial enterprise. The business houses and public institutions comprise four hotels, three drug, one jewelry, one boot and shoe, two millinery and four general stores, three wagon and blacksmith shops, a furniture factory, a hardware and farm machinery establishment, a grist mill, livery stables and a bank. The village has churches of the various denominations, most of which have fine edifices, and also one of the best graded public schools in the State, with high school department, conducted by an able corps of teachers.

Prominent among the features of this section are its advantages as a summer resort. *Detroit Lake*, one of the most beautiful sheets of water in Minnesota, lies only half a mile from the business portion of the village. Each year it becomes more popular with the people of the neighboring towns, and also with those who are accustomed to flee from the hot and dusty cities during the summer months. The lake, which is about a mile and a half wide, and seven miles long, in form somewhat resembles a horseshoe, with a sand bar reaching from shore to shore, about midway between the two ends of the lake, which is converted into a most delightful driveway. Here is a high bank towering above the clear waters of the lake, and there the broad and pebbly beach, with an occasional "opening," where a sturdy frontiersman is carving out a farm. To the east, Detroit Mountain, whose heights are covered by a dense growth of timber, towers far above the surrounding country, lending its rugged charms to the scene. The lake is stocked with all kinds of "gamey" fish, which are an attraction to the sportsmen, the variety including pickerel, black and Oswego bass, wall-eyed pike, perch, and also California salmon, which were planted in the lake some time ago by the State Fish Commissioner.

In 1882 the Detroit Lake and St. Louis Boat Club organization was perfected, and its members purchased and improved a handsome piece of property fronting on the east bank of the lake. They have now a fine and commodious club house. The club is limited to one hundred members, each of whom purposes to erect a handsome private summer residence upon the grounds in the near future. Other gentlemen have purchased lots fronting on the south bank of the lake for the purpose of building cottages.

Detroit Lake, however, is only one of many which abound in the immediate vicinity, the following being also within the township, and varying from one to four miles in length, viz., *Floyd Lake, Lake Flora, Lake Rice, Oak Lake, Edgerton Lake, Long Lake* and *Lake St. Clair*. Here, too, are mineral springs, iron and sulphur, the health-giving qualities of which have been known to the Indians for many generations.

The Detroit Lake Pleasure Grounds are the most popular place of amusement in northern Minnesota, and are to be made more than ever attractive. A handsome steam yacht, as well as sail and row boats, are furnished on these grounds to visitors at a small cost. A new hotel, costing $25,000 or more, has been erected, and several other new buildings that will be ornaments to the town are proposed. Next year a court-house is to be built at a cost of $25,000, and this will establish permanently the seat of justice here. A grain elevator of 30,000 bushels capacity is also in process of construction.

Looking upon Detroit to-day, with its evidences of advanced civilization, it seems almost incredible that only twelve years ago the pioneer settlers were met here by a band of Chippeways, who, recognizing and graciously succumbing to the inevitable march of events, invited them to a feast of fat things, *viz.*, baked dog and boiled fish, cooked whole, with entrails included. Of these Indian dainties the pioneers partook, but with what degree of relish has never been recorded. Tri-weekly stages run from Detroit to *Richwood*, *White Earth*, *Cormorant*, *Spring Creek*, *Pelican Rapids* and *Carsonville*, which are also favorable points to visit in search of feathered game, and also for bear and deer. The latter are met along the woody margins of streams and lakes, while Bruin confines himself mainly to the coppices and forests.

An Effective Indian Tableau.—Twenty-five miles north of this village is the White Earth Reservation of the Chippeway Indians. These Indians, who call themselves Ojibways, have always been the friend of the white man. They were a kindly disposed race, and contact with white men had dragged them down into a depth of degradation never known to their fathers. The deadly fire-water flowed throughout their country, and disease, poverty and death held a carnival in every Indian village. Their friends secured for them this beautiful reservation, as fair a country as the sun ever shone upon. This action might have been prevented by the pioneers of the Northern Pacific Railroad; but in this case, as in every other where the rights of the red men were concerned, the railroad company was his friend. A few years after Bishop Whipple had commenced his mission here, the treasurer of the company, the Bishop, Lord Charles Hervey and others paid the Indians a visit. The Bishop consecrated their hospital and held confirmation. After the services the Indians made a feast for the Bishop and his friends. When all had eaten, the chief, Wah-bon-a-quot arose, and, addressing the Bishop, said: " We are glad to see our friends. Do they know the history of the Ojibways? If not, I will tell them." In a few graphic words he described the Indians as they were before the white men came. The woods and prairies were full of game, the lakes and forests with fish, and the wild rice brought its harvest. "Hunger never came to our wigwam," said he. "Would your friends like to see us as we were before the white man came?" Suddenly there appeared a tall athletic Indian, with painted face, and dressed in a robe of skins ornamented with porcupine quills; and by his side a pleasant-faced woman in wild dress. "There," said the chief, with eyes gleaming with pride, "there, see Ojibways as they were before the white man came." Turning to his guests, continued he, "Shall I tell you what the white man did for us?" Then dropping his voice, he added, "The white man told us we were poor; we had no books, no fine horses, no fine canoes, no tools. 'Give us your land, and you shall become like the white man.' I cannot tell the story; you must see it." Then stepped out a poor, ragged wretch, with tattered blanket, and face covered with mud; by his side a more dreadful specimen of womanhood. The chief raised his hands: "Are you an Ojibway?" The Indian nodded. Sadly the chief said: "Oh, Manitou, how came this?" The Indian raised a black bottle, and spoke one word, "Ishkotah wabo" (fire-water). "This is the gift of the white man." It went like an electric thrill through every heart, and brought tears to many eyes. The chief said: "A pale-faced man came to see us. I am sorry to say he has seen me and my fellows drunk. He told a wonderful story of the Son of the Great Spirit coming to save men. He told us his fathers were wild men; that this religion had made them great, and what it had done for them it would do for others. We did not hear; our ears were deaf; our hearts were heavy. He came again and again, always telling one story of Jesus, the poor man's friend. We knew each summer that when the sun was high in the heavens the Bishop would come. He gave us a red minister. At last we heard. Shall I tell you what this religion has done for my people? You must see." There stepped out a young Indian in a black frock-coat; by his side a woman neatly clad in a black alpaca dress. "There,' said the chief, "there is only one religion which can take a man in the mire by the hand and bid him look up and call God his Father."

There are 1,500 civilized Indians at White Earth. They have two churches—the Episcopal church, under the care of Rev. J. A. Gilfillan (white), and Rev. J. S. Enmegahbowh (Indian), and a Roman Catholic church. Visitors are always received with kindness, and no excursion on the line of the Northern Pacific Railroad will be more pleasant than a visit to White Earth.

Minnesota Division. 79

Audubon—(234 miles from St. Paul; population, 170.)—This settlement, in Becker County, is principally of Scandinavians. It was founded about 1872, and named after the celebrated naturalist. It has had a slow but steady growth, being a good point for the production of wheat, oats, barley, rye, potatoes, butter, cheese and eggs. Audubon has four general stores, two churches, two wheat elevators, a grist mill, a saw mill and the usual shops. There are several lakes in the vicinity, which afford good fishing, and small game also abounds.

Lake Park—(240 miles from St. Paul; population, 500.)—This is a young but thriving business town, in Becker County, thirty-four miles east of Fargo. It has three grain elevators, and a fourth, and larger one, is in course of construction. There are also real estate offices, a bank, a newspaper, several dry goods stores, hardware and drug stores, a church, a public hall, a flouring mill, and telephone communication between depot and bank. The large farms of Thos. H. Canfield and M. E. d'Engelronner are in the neighborhood. Mr. Canfield has nine and a half sections, or 6,080 acres, most of which is under cultivation, affording employment to a large force of men and teams. The principal production at present is wheat, but the raising of blooded stock is also extensively engaged in. Lake Park is situated on a lovely lake, in what is known as the Park Region. The soil is a rich black loam. The country is dotted with lakes, which are well stocked with fish, and large numbers of deer and other game are killed in the vicinity for market. The town has a summer hotel, accommodating a hundred people. Twenty-two miles northwest of Lake Park is the White Earth Indian Reservation, a pleasant place to view the manners and customs of the red men, who are on friendly terms with the whites. They have farms under a good state of cultivation. *Hillsdale* (two miles beyond) is a side track.

Hawley—(251 miles from St. Paul; population, 200.)—The situation of Hawley is quite attractive. The town lies in the depression east of the hills which skirt the Red River. It is supplied with a good school and two churches, one of which is Methodist, and the other belongs to the United Brethren. From the town, the distance is but a few minutes' walk to the Buffalo River, where there are two large flouring mills. The business blocks of Hawley are commodious. All branches of trade are represented and flourishing. The town has one hotel, an opera house and an elevator.

Silver Lake, three miles south, a beautiful body of water covering 300 acres, is an excellent fishing resort. Good hunting and fishing are also to be had in the surrounding country, geese, ducks and grouse being quite plentiful, while deer and bear are found in the timber regions southward.

Muskoda—(256 miles west of St. Paul; population, 125.)—Muskoda is an Indian word, said to signify "the buffalo river." The Buffalo River runs adjacent to the town, and is a beautiful swiftly flowing stream, fifty feet wide, with high timbered bluffs on either side. It is well adapted to milling purposes, and abounds in black bass, pike and pickerel. *Lake Maria*, two and a half miles southeast of Muskoda, and half a mile south of the Northern Pacific track, is a curiosity in itself, inasmuch as it is not known to contain a living thing, although every other lake in the region is full of fish. This lake covers 300 acres, and is twelve to fifteen feet deep. A beautiful forest surrounds it, and its shores are a gravelly beach. *Horse-shoe Lake*, two and a half miles north of the Northern Pacific Railroad, covers 200 acres, and is well stocked with fish.

The soil of the surrounding country is rich and well adapted to the production of cereals and grasses, the region being noted for wheat and stock raising. There are a number of boiling springs here, from which as pure water flows the year round as can possibly be found anywhere. This neighborhood is also famous for its abundance of small game ; geese, ducks, prairie chickens, snipe and rabbits being among the varieties. In former years the country was a favorite hunting ground of the Indians, and the region is strewn still with buffalo skulls and elk horns.

Muskoda is the point where a remarkable English colony settled in the spring of 1872. It consisted of seventy-five persons, among whom were ten ministers, seven music teachers, four lawyers, fifteen school teachers, five doctors, two dancing masters, twenty clerks, five farmers, two wagon makers and five blacksmiths. The farmers, wagon makers and blacksmiths are still at Muskoda, and doing well, but the sixty-three others have long since disappeared. The town has at present a flouring mill, which has more work than it can do; a convenient depot, grain warehouse and elevator, a general store, blacksmith and carpenter shop, and also a church and a school-house. The most extensive sand and gravel bank in the Northwest is found here, whence 30,000 cars of gravel and sand were shipped to Fargo and Moorhead alone, for the improvement of those towns, during the summer of 1882.

Glyndon—(264 miles west of St. Paul; population, 450.)—Glyndon lies in Clay County, four miles west of the Northern Pacific crossing of the North Buffalo River, and nine miles east of the Red River of the North. The town was founded in 1872, by the location here of the crossings of the Northern Pacific Railroad and the St. Paul, Minneapolis and Manitoba Railroad. Here were the field head-quarters of the Red River colony of the date named. Though set on a level prairie, Glyndon possesses some picturesqueness from its situation between the two branches of the Buffalo River, which flow to the west and north, and it shows the activities peculiar to the crossing town of two great railways. The fact that it is in ready communication with the East and South by rail and lake routes, makes it a superior wheat market and shipping point. The present vast business of grain buying and warehousing was begun at Glyndon, and here was built the first grain elevator of the great Northwest. The Barnes and Tenny farm, 4,000 acres in extent, is still one of the features of the locality, affording a specimen of the rich and productive agricultural lands which surround the town. Drainage is good by streams and coulées. In the village are six stores, lumber yard, three machinery depots, three hotels, two blacksmith shops, one Union and one Methodist church, graded schools, a large public hall, suitable for dramatic or musical entertainments, a printing office and weekly newspaper, a 5,000 bushel grain elevator, and extensive railway building and yard facilities, including several miles of side tracks. Wheat raising is still the leading farm industry, but the stock and dairy interest is growing rapidly, the excellent natural grasses of the region building up a flattering amount of traffic in milk, which is shipped by rail West and North. The population of Glyndon is largely composed of Americans. *Tenny*, three miles further west, is a side track, with an elevator for the storage and shipment of wheat.

The Red River of the North.—This stream is named to distinguish it from the Red River of Louisiana. It rises in Lake Traverse (lat. 46°), and after meandering southward among the lakes of Minnesota, flows due north a distance of more than 200 miles, entering Lake Winnipeg, in the northern part of the Province of Manitoba. The Red River marks the boundary between Minnesota and Dakota. Its elevation above the sea level at Moorhead and Fargo is 807 feet. From these points northward to Winnipeg the stream is navigable, even at a low stage of water, the shallow portions being dredged as occasion requires. Large quantities of wheat and merchandise are transported by steamers which ply between Moorhead, Fargo and

Winnipeg. In 1882 the fleet numbered sixteen steamers, of a capacity of from 100 to 250 tons each, and twenty-one barges of thirty tons each.

This river is always subject to overflow in the spring. Its course being almost due north, the winter ice breaks up first along its southern length, and the frozen stream cannot carry off the freed waters, which back up upon the ice and deluge the fields to a greater or less extent. There can be no question but that the soil is benefited by the alluvial deposits which are thus spread over it, but it is often very inconvenient and discouraging to the settlers in Manitoba to be cut off from rail communication with the outer world by the overflow. The valley of the Red River of the North is from sixty to eighty miles wide, embracing an area of 67,000 square miles, at least eighty per cent. of which is composed of the very best farming land. The valley proper is a beautiful prairie, apparently as level as a garden bed, though, in reality, sloping gently and imperceptibly from both sides to the river, and slightly inclining to the north. The soil consists of a rich black loam, from three to seven feet in depth, which yields from twenty to twenty-five bushels of wheat per acre. The whole valley is well watered by Nature, there being a large number of small rivers, tributary to the Red, on either side, which perform the double office of supplying water and draining the land. The most important of these streams on the Minnesota side are, the Buffalo, Wild Rice, Marsh, Sand Hill, Red Lake, Middle, Tamarac, Two Rivers and Red Grass. From the west there are several rivers of considerable size, the principal being the Cheyenne, Goose, Turtle, Forest, Park, Tongue and Pembina. All of these have branches, which penetrate the level prairie in every direction, affording an abundance of excellent pure water. The rivers are, for the most part, skirted with a good growth of oak, elm, soft maple, ba-swood, ash and box elder, which is ample for fuel purposes. Extensive pine lands are about the head waters of most of the rivers on the Minnesota side.

On examining a map of the Red River Basin, the fact is apparent that most of the tributary streams have their sources in a higher latitude than their mouths. This peculiarity extends as far north as the Saskatchewan, in Manitoba, and suggests that, originally, the slope of the country was to the south, and that the waters of this immense area were drained by a large stream, which occupied the now comparatively dry valley of the Minnesota. The theory has been advanced by scientific men, that there has been a subsidence along the valley of the Red River, having its maximum below Lake Winnipeg, together with a possible upheaval at the head waters of the Minnesota River.

Moorhead—(273 miles northwest of St. Paul; population, 4,000.)—This sprightly city, in lat. 46° 51′ N., long. 96° 50′ W., and 840 feet above the level of the sea, is the last place on the line of the Northern Pacific Railroad in the State of Minnesota, distant 251 miles from Duluth, on Lake Superior. It is the county-seat of Clay County, advantageously situated on the east side of the Red River of the North, immediately opposite the bustling city of Fargo, Dak., with which it is in communication by means of a bridge which spans the stream. Being situated in the midst of the great wheat region of the Northwest, its growth has been steady and substantial, and its financial solidity is unquestionable. Moorhead has fine business blocks, flouring mills, grain elevators, a brewery, a driving park, fair grounds, a daily and weekly newspaper. Its chief hotel—the Grand Pacific—perhaps the largest and best equipped hotel in the Northwest, was built at a cost of $160,000. Its architecture is in pure Queen Anne style, the interior fittings and decorations being in keeping. In addition to this, there is another first-class brick hotel, three stories in height, with accommodation for eighty guests. Moorhead schools afford superior advantages. Besides the public schools, there is a flourishing academy under the control of the Episcopal church, which is known as the "Bishop Whipple School," in honor of the respected Bishop of the diocese of Minnesota, and this establishment offers a classical as well as business education. The churches, represented by all the leading denominations, have commodious edifices. A number of miscellaneous manufacturing enterprises already exist, among which may be named an iron foundry and a planing mill, and other important industries are to be established. The city is well supplied with an abundance of brick, there being four yards in which this building material is manufactured, giving employment alone to 150 men.

Moorhead is the crossing point of two trunk railroads, the Northern Pacific and the St. Paul, Minneapolis and Manitoba. Besides these two great railways, there are also the Moorhead and Northern, from Moorhead to Fisher's Landing, Minn., and the Moorhead and McCauleyville, from Moorhead to McCauleyville, Minn. The principal product of the country is wheat, and large shipments of the same are made, not only by rail, but also by river. One of the steamboat lines—the Alsop—owned at Moorhead, does a heavy freight business, towing barges laden with supplies and produce to all points on the Red River between Moorhead and Winnipeg, Manitoba.

Northern Dakota.

The railroad crosses Dakota from east to west in nearly a direct line, its length within the Territory being 294 miles. The entire area of Dakota is 153,000 square miles, and it is only exceeded in size by Texas and California. In 1870 the total population of Dakota was 14,000, and for many succeeding years the progress of settlement was quite slow. Since 1879, however, the inflow of people, especially into the wheat lands in the northern part of the Territory, has been extraordinary. It is estimated by competent authorities that the number of inhabitants in the spring of 1883 amounted to 325,000, and that the great majority of the new comers were attracted by the profits of wheat raising. The mass of immigration has settled upon the prairie lands which stretch out, with little interruption, for a distance of fifty miles on either side of the Northern Pacific Railroad, the entire length of the Territory. The glory of this great belt of country is its fertile soil and a climate perfectly adapted to the production of cereals. This region already plays an important part in the wheat-growing area of the United States, a yield of twenty bushels per acre being usual and twenty-five bushels not an extraordinary crop. The general character of the land is that of a rolling prairie, interspersed with broken butte formations west of the Missouri River. The entire country is fairly watered by the Red River of the North, the Cheyenne, the Dakota or James, the Missouri and other streams, with their many tributaries, as well as by numerous lakes in the northern and eastern portions, some of which are of great size and beauty. Good well water is everywhere found by digging to a reasonable depth.

In 1880, just six years after the capacity of the soil was first tested in the Valley of the Red River, the yield of wheat along the line of the railroad was about 3,000,000. In 1881, so great was the increased acreage, there was a product of 9,000,000, and in 1882 the crop was 12,000,000. In coming years the quantity of new land which will be put under cultivation bids fair even to be in a larger ratio than that which has marked the increasing acre-

Sarah Carman
Belleville

Elijah J. Flint
Belleville Ont.

Wm Watt
Brantford. Ont.

Thos. Brossoit
Beauharnois

H. J. Matheson
Perth

Arthur J. Matheson
Perth Ont.

age since 1879. The wheat of northern Dakota has no equal for milling purposes. It is preferred by the great millers at Minneapolis and elsewhere throughout the United States to any other variety, being best adapted to the modern methods of making flour. It is raised from seed known as Scotch Fife, and is graded in market as "No. 1, hard," bringing an excess of ten or fifteen cents per bushel over the soft varieties. Under the new process of manufacture it has been demonstrated that flour produced from hard spring wheat is a far more profitable commodity than that made from winter wheat. For example, bakers are able to get 250 pounds of bread from a barrel of flour made from the hard spring wheat, and only 225 pounds from the same quantity of flour which is ground out of winter wheat.

Prairie Farming.—The cultivation of the soil in a prairie country is, in some of its processes, very different from the methods pursued elsewhere, and has given rise to at least two new technical terms, which are known as "breaking" and "backsetting." Premising that the prairie soil is free from roots, vines or other obstructions, and that the virgin sod is turned from the mould board like a roll of ribbon from one end of a field to the other, a fact is presented which farmers who are accustomed to plow among stones, stumps and roots can scarcely grasp. But the sod thus turned is so knit together by the sturdy rootlets of the rank prairie grass, that a clod of large size will not fall apart even though it be suspended in mid air. To "break" or plow this mat, therefore, it is necessary to cut it, not only at the width of the furrow it is desired to turn, but underneath the sod at any thickness or depth as well. An ordinary plow could not endure the strain of breaking prairie soil, so plows called breakers have been constructed to do this special work. Usually, three horses abreast are employed, with a thin steel, circular coulter, commonly called a "rolling coulter," to distinguish it from the old-fashioned stationary coulter, beveled and sharpened for a few inches above the point of the plow to which it is attached. A furrow is broken sixteen inches wide and three inches thick, and the sod, as a rule, is completely reversed or turned over. Each team is expected to break sixteen miles of sod, sixteen inches wide and three inches thick, for a day's task. By cutting the sod only three inches thick, the roots of the grasses, under the action of heat and moisture, rapidly decay. The breaking season begins about the 1st of May and ends about the 1st of July. The wages of men employed at this kind of work are $20 per month and board. The estimated cost of breaking is $2.75 per acre, which includes a proportionate outlay for implements, labor and supplies. But the ground once broken is ready for continued cultivation and is regarded as having added the cost of the work to its permanent value. The "broken" land is now with propriety termed a farm.

"Backsetting" begins about the 1st of July, just after breaking is finished, or immediately after the grass becomes too high, or the sod too dry to continue breaking with profit. This process consists in following the furrows of the breaking, and turning the sod back, with about three inches of the soil. In doing this work, it is usual to begin where the breaking was begun, and where the sod has become disintegrated, and the vegetation practically decomposed. Each plow, worked by two horses or mules, will "backset" about two and a half acres per day, turning furrows the width of the sod. The plows have a rolling coulter, in order that the furrows may be uniform and clean, whether the sods have grown together at their edges or not. The "backsetting" having been done, there only remains one other operation to fit the new ground for the next season's crop. This is cross-plowing (plowing crosswise, or across the breaking or backsetting), or so-called fall-plowing, which is entered upon as soon as the threshing is over, or on damp days during the threshing season. A team of two mules will accomplish as much cross-plowing in a day as was done in backsetting — two and a half acres. The wages for backsetting and fall-plowing are also $20 per month and board, or $1.50 per acre to hire the work done.

The virgin soil having been broken, backset and cross-plowed, is now ready for seeding. This, ordinarily, begins from about the middle of March to the 1st of April, and is often not finished until the 1st of May. Instead of the old style of hand sowing, a broadcast seeder is used, one of which machines will sow twelve acres a day. Fifty-two quarts of clean Scotch Fife seed wheat are used to the acre. The cost of sowing the ground is seventy-five cents per acre, and the average cost of the seed wheat, upon the larger farms, has been $1.50 per acre. Seeding having been carefully attended to, the harrowing, or covering process, demands close attention. The grain must be evenly covered, at a uniform depth, to ensure a good stand, healthy growth and even maturity. On the so-called bonanza and systematically conducted farms, one pair of harrows follows each seeder, going over the ground from one to five times, according to the condition of the soil, until it is well pulverized, the seed evenly covered, and the surface reasonably smooth.

Harvesting on the large farms begins about the 1st of August. Self-binding harvesters, one to every 160 acres, are employed, and one driver and two shockers are required to each machine. The wages during the harvest season are $2 per day and board.

J. Gregory
Napanee
Ont.

George Lewis
Toronto.

J. Krug
Berlin.

Emilie Bryson
Beauharnois
Que.

Seeding on a Bonanza Farm.
[By permission of Harper & Brothers, New York.]

The work on a wheat farm only occupies a few weeks in the year, and the business is attractive on that account, apart from the profits. After the plowing and seeding are finished, the farmer can look on, and see Nature grow and ripen his crop, until the harvest-time comes. By the end of August the year's work is practically done. Expensive farm buildings are not required, for the grain may be threshed in the fields, and hauled immediately to the nearest railroad station. Very little fencing is needed on a wheat farm. Frequently the cultivated portion is left unenclosed, and a barbed wire fence is put around the pasture lot to secure the cattle. The outlay for improvements is comparatively light; and, as the country is open and ready for the plow, the settler makes a crop the first year, and is tolerably independent from the start. A village, with school house, post-office, stores and churches, springs up, as if by magic, in the neighborhood of his home, and he suffers few of the privations which used to attend frontier life.

The extent of the northwestern wheat region cannot now be estimated, nor its future productiveness foreseen. It includes nearly the whole of Dakota, east of the Missouri River, and a considerable portion of the western half of the Territory. The wheat-growing industry has been steadily moving west for more than half a century, and the rich lands of the Red River Valley of the North, and the vast rolling plains of Dakota and the Pacific Northwest, must, ultimately, be the permanent wheat field of the continent.

Cost of Farming New Land.—Settlers on the line of the railroad have the option of taking 320 acres of the public lands, by complying with the liberal terms of the Homestead Law and the Tree Culture Act, or of buying good agricultural land, on easy terms, from the Railroad Company. In either case, the cost of opening a farm is the same, and the expense of preparing prairie soil is:

Breaking	$2 75 per acre.
Backsetting.	1 50 per acre.
Seed (taking one year with another)..........	1 50 per acre.
Putting in crop	1 00 per acre.
Cutting, binding and shocking	2 00 per acre.
Threshing and marketing	2 50 per acre.
	$11 25 per acre.

The cost of a crop from stubble ground, after the farm is opened, in the second and succeeding years, would be as follows:

Fall plowing	$1 75 per acre.
Seed wheat	1 50 per acre.
Putting in crop	50 per acre.
Cutting, binding and shocking	2 00 per acre.
Threshing and marketing	2 50 per acre.
	$8 25 per acre.

These estimates are on the basis of hiring the labor and machinery. If a farmer owns his own team and implements, he can reduce the cost about $2 per acre. The expense of the buildings, teams, machinery and household effects necessary to open wheat lands and keep them under cultivation, is $10 per acre, and this is called the permanent working capital. From this it is evident that the outlay for raising the first crop on a prairie farm is $20 per acre.

In case the first yield is twenty-four bushels per acre, which is the usual average, and the wheat is sold for ninety cents per bushel, which is the ordinary price, the farmer is reimbursed for his buildings and equipment and the expense of raising his first crop. These estimates may be thus summarized:

Cost of preparing land for farming	$11 25 per acre.
Less, in case of a farmer owning his own teams and machinery	2 00 per acre.
	$ 9 25 per acre.
Buildings and equipment	10 00 per acre.
	$19 25 per acre.
Allow for interest, etc	2 35 per acre.
	$21 60 per acre
Realized from first crop, twenty-four bushels, at ninety cents per bushel	$21 60 per acre.

The ordinary farmer of 160 acres generally puts about $3 per acre into a house, $2 per acre in a stable, and provides himself with two spans of mules or horses, one gang plow, one seeder, two pairs of harrows, one mowing machine, one self-binder and one wagon, hiring an itinerant thresher at a fixed price per bushel. The new comer usually does not care to break up his entire 160 acres the first year, but gets his farm in condition gradually, working part of the time for his older neighbors. In this way he earns a living for himself and family until his own crop is harvested.

The agricultural products include the whole range of those common to the Northern States. Oats and barley yield largely, the former running from forty to sixty bushels per acre, and selling for forty-five cents per bushel, and the latter for seventy-five cents. Dairying is not carried on to a great extent, because wheat growing is more profitable. The country, however, is well adapted to dairy farms, as the native grasses of northern Dakota, particularly the blue joint and high prairie grass, are as nutritious as the cultivated grasses of the Middle and Eastern States.

One of the principal factors in profitable wheat culture is easy and cheap transportation. The farmer of northern Dakota is amply provided for in this respect. He has the choice of two outlets for his grain and other products. It is only 250 miles from the Red River to Lake Superior, whence wheat is shipped *via* Duluth and Superior City to the markets of Buffalo and New York, while the immense mills at the Falls of St. Anthony, in Minneapolis, create a demand which has never yet been fully satisfied. The uniform rate of freight for carrying wheat adopted by the railroad gives every shipping point on the line equal advantages in the cost of getting grain to market.

Prairie Soil and its Constituents.—From an essay of Dr. Charles Louis Fleischmann, of Washington, a recognized authority in scientific agricultural affairs, the following interesting extracts are made:

"If nature had not stored up in the far West immense tracts of inexhaustible masses of vegetable mould, enabling farmers to keep up their lands in a high state of fertility, the ruinous system which has well-nigh worn out the lands of the Eastern States would be continued until every remaining acre would become exhausted.

"This extraordinary accumulation of vegetable mould extends over a large part of the Northwest. Some of the States possess more prairies than others. Iowa, for instance, is one-third prairie land, and the northern part of Minnesota is almost a continual prairie.

"We may assume that there are at least 100,000,000 acres of prairie lands which contain all the elements as well as the inorganic substances which plants require for a perfect development. This valuable accumulation of mould varies from two to six feet in thickness, and often even more. It is free from all admixture of earthy substances, such as clay, sand, etc., and, in a dry state, the prairie soil has no cohesion, but crumbles to dust. It absorbs water very rapidly, and loses it equally as fast. When wet its color is coal black; when dry it turns gray. In the upper layer it contains some carbon and burns like very poor peat, leaving a large amount of ashes. When the prairie soil is exposed to the blast of a forge it melts and backs together like slag, consisting mostly of silicate of potash. Like all decayed vegetable matter it contains a large amount of ammonia.

"The prairies of the Northwest were once lakes, some of which were of considerable extent. As the rivers cut deeper channels these lakes were gradually drained of their waters. In their beds sprung up aquatic plants, and, after many, many centuries, large accumulations of vegetable mould were deposited. Finally, when the lakes were completely drained, the vegetation changed, and upon the nymphaceous remains more nutritious plants sprung up and formed pastures for buffaloes and other herbivorous animals. According to this theory, the first layer of vegetable mould must contain a large amount of carbon, because the aquatic plants, having been protected by the water against fire, would have carried their entire carbon into their watery grave. From this first layer of decayed plants the succeeding vegetation must have drawn its nutriment of inorganic substances, as the roots could not have passed through the whole thickness of the vegetable mould to seek food in the mineral subsoil of the bed of the lake; but the original accumulated carbon was left. According to this view the upper layer or surface stratum contains the original inorganic substances which were taken from the mineral subsoil. Consequently, the productiveness of the prairies and their durability for producing crops depend on the thickness of the surface layer.

"This shows how important an accurate examination of the prairie soil is to the farmers of these regions.

"How long it took to produce that enormous mass of vegetable mould can only be conjectured. Let it be assumed that a crop of one year from an acre of prairie amounts to 2,000 pounds of dry grass, yielding about 130 pounds of inorganic substances and on an average of 1,500 pounds of carbon. When these 1,630 pounds of decayed vegetable matter are scattered over one acre of 43,560 square feet, every square foot of land would receive a delicate film of ashes, and it would require at least 500 such films of ashes to produce one inch, or 6,000 to make one foot in thickness. Consequently it must have taken about 36,000 years to produce six feet in thickness of vegetable mould.

"The ash constituents of grasses differ very little in the various species, so that they result in no great difference in the fertility of the prairie soils. A crop of dry grass, weighing 2,000 pounds, yields on an average 130 pounds of ashes, which contain: potash, 34; soda, 9.5; magnesia, 6.6; lime, 15.5; phosphoric acid, 8.2; sulphuric acid, 6.8; silica, 39.3; chlorides, 10.6, and sulphur, 2.4. This would be the average contents of ash constituents in the prairie soil, giving all the substances which the cereals require.

"When a crop of wheat or corn is taken from that soil, provided the straw is returned to the soil in any shape whatever, the crop consumes only a portion of the plant constituents of a single hay crop, with the exception of the phosphoric acid, which is entirely extracted by a wheat crop from the ashes of a hay crop. For example, 1,000 pounds of wheat yield 17.7 parts of ashes, which contain 5.5 potash, 0.6 soda, 2.2 magnesia, 0.6 lime, 8.2 phosphoric acid, 0.4 sulphuric acid, 0.3 silica and 1.5 sulphur.

L. J. Pinault
Québec

J. P. Roy, M.D.
Québec

Lattie E. H. Hawbert
Toronto

Harrowing on a Bonanza Farm.
[By permission of Harper & Brothers, New York.]

"According to the above, one layer of the ashes of hay, of the thickness of finest paper, spread over an acre, would produce a wheat crop of twenty bushels. Therefore, a thickness of one inch of prairie soil would furnish 500 wheat crops. Speculations on paper do not always agree, however, with practical experience. So much is sure, nevertheless, that the prairie soil is exceedingly rich in the ash constituents of plants, and it will serve for a long time for the production of cereals.

"Yet prairie farmers must bear in mind that the vegetable mould is of a different nature than the earthy or mineral soils, like clay or sand. The prairie soil contains only a certain amount of the inorganic substances necessary to the growth and perfect development of plants. When one of the inorganic constituents is exhausted, it cannot be replaced by fallowing or mechanical means, as is the case with the mineral soils, which by the disintegration of the coarse, earthy substances replace again the lacking constituent.

"When the straw of wheat crops is not burnt, and the ashes returned to the soil, the prairie soil will lose silica, and the succeeding crops will show a certain weakness in the stem of the straw.

"In view of the formation, extent, richness and importance of the vegetable mould of our Northwestern prairies, it is established to a certainty that the United States is in possession of one of the greatest treasures in existence, which is not surpassed in value and importance by all the precious metals in the bowels of the earth. If only one foot in depth of prairie soil were set aside for manuring purposes, leaving two-thirds of that soil for future cultivation of the prairie region, the portion destined for manure would amount in round numbers to one hundred and thirty billions of tons. This quantity would be large enough to restore all our exhausted soil, besides improving the mineral lands and meadows of the Northwest, thus enabling the farmer not only to raise all the breadstuffs and meat for our rapidly increasing population, for all time to come, but to assist other nations in case of need.

"If farmers do not burn the wheat straw, and faithfully return the ashes to the field, there must necessarily be a falling off in the fertility of the soil. The straw requires, for each 2,000 pounds in weight, fifty-six pounds of silica—sixteen pounds more than a hay crop would furnish. The soil, therefore, is not able, after many wheat crops have been taken away, to provide the straw with sufficient silica, and its ability to support the ear is lessened, causing the stalk to lodge, and producing an imperfect crop. This being the case, the lacking silica must be furnished by the addition of a mineral soil, in order to make the prairie soil compact, and prevent the wheat straw from lodging. An addition of pure sand would even subserve this purpose."

Philological and Historical.—Dakota is named after the great Indian nation who once claimed a large portion of the Northwest for their own. The northern Indians are divided into two great families: the Algonquins, which include the Chippeways, or Ojibways, the Ottawas, the Crees and a host of others, and the Dakotas or Sioux, who are divided into many smaller bodies, all speaking the Dakota language. The only difference is, that the Dakotas east of the Missouri use a D, where those west use an L. For example, those east say: "codah," "friend"; those west, "colah," "friend." Those east call themselves Dakotas; those west, Lakotas. The Lissetons, Wahpetons and Mandawatons, who lived in Minnesota, were called Santees. The Yanktons, Yanctonais, Brulé, Cuthead, Ogallas, Two Kettles, and a score of other bands, are Sioux. Nicolet, Catlin and others say that they are one of the finest specimens of wild men on the earth. For a generation they were our devoted friends. Our first fight with the Sioux was near Fort Laramie. Some Mormons, who were crossing the plains to Utah, had a lame ox, which they turned loose to die, and a camp of Indians found and killed it, and made a feast. The Mormons saw this in the distance, and, thinking they could secure payment, stopped at Fort Laramie, and told the officer in command the Indians had stolen their ox. The officer, who was half drunk, took some soldiers, went to the Indian village, and demanded the ox. The Indians said: "We thought the white men had turned him loose to die. We have eaten the ox; if the white men want pay for him, you shall have it out of our next annuity." "No," said the drunken officer; "I want the ox, and, if you do not return him, I will fire upon you." He did fire on them, and killed a chief. The Indians rallied and exterminated the command. That war cost one million of dollars.

Fargo and Southwestern Branch Railroad.

FROM FARGO TO LISBON. DISTANCE, 56 MILES.

This road is opening to settlement a fine agricultural region in central Dakota. It is now operated to the thriving town of Lisbon, fifty-six miles from Fargo, and will soon reach La Moure, county-seat of the county of the same name, thirty-two miles beyond. Passing by the newly established towns and stations of *Cotters, Horace, Davenport* and *Sheldon*, distant, respectively, four, ten, nineteen and forty-one miles from Fargo, each having a substantial and increasing tributary agricultural business, the important town of Lisbon is reached.

Lisbon—(56 miles from Fargo; population, 1,000.)—This is the present terminus of the railroad. The village is situated very pleasantly on the Cheyenne River, being sheltered by forests and towering bluffs. Lisbon was first started in 1881, when few people had settled in Ransom County, and although for some time it had no railroad facilities nearer than thirty-five miles, its growth has been remarkable. From a mere speck in the valley it has risen to a thriving city of 1,000 inhabitants, with a full city government. Educational interests have been well looked after by a competent school board. A fine two-story school building, with four rooms, has lately been finished. There are three church organizations, namely: Presbyterian, Methodist and Baptist, and three weekly newspapers. There are grain warehouses, an elevator, banks, a brick yard, and all the stores and shops which are needed to carry on the large trade of the rich agricultural region of which Lisbon is the centre. The soil, for at least fifty miles in every direction from Lisbon, is a black, sandy loam, with a clay subsoil, and for the production of wheat, oats, barley, flax, peas, root crops and vegetables generally is not excelled. The average product per acre of wheat is twenty-two bushels, of oats fifty bushels, while forty-five to fifty bushels of corn, fully matured and ripened, is not an unusual yield. A flouring mill, driven by water power, is in successful operation, and several other water powers near the city invite the establishment of other manufacturing enterprises.

An Indian Spartan.—After the Sioux massacre of 1862 Gen. H. H. Sibley, with singular wisdom, enlisted a body of Sioux scouts under their chief, Gabriel Renville. These scouts were placed in camps twenty miles apart, and ordered to kill any Indians who penetrated the country to commit murder. One party did enter; all were killed but one young warrior, who was brought in a prisoner. His own uncle was in command of the camp near where Moorhead now stands. As he entered he saw his uncle, and reached out his hand, saying: "I am glad to see you, my uncle; you will save my life." His uncle said: "I am a soldier, my nephew; my orders are to kill any man who has blood on his hands. Your hands are stained with white blood—you must die." He took up his gun and killed him.

Dakota Division.

FARGO TO MANDAN.—DISTANCE, 200 MILES.

Fargo—(274 miles west of St. Paul; population, 10,000.)—This city, the county-seat of Cass County, Dakota, 242 miles west of Lake Superior, is situated on the western bank of the Red River, which, though a very tortuous stream, is the constituted boundary line between the State of Minnesota and Dakota Territory. Nine or ten years ago the place was a mere hamlet, and business lots that are now worth from $1,000 to $5,000 each could have been bought for a song. Fargo did not begin to grow until after Mr. Dalrymple had opened his immense farm in the vicinity, and proved, by actual experiment, the remarkable fertility and great agricultural value of the Red River Valley. From that time till the present the growth of Fargo has been rapid and the increase in business almost marvelous. The population since 1880 has nearly doubled every twelve months, and increase in almost the same ratio is probable for years to come. The people are enterprising and ambitious, striving vigorously to promote the interests of the place, and zealous to maintain the position which has been so rapidly achieved. Fargo is the very liveliest type of a new Western town, with all the modern improvements, including street cars and electric lights. Many important manufactories have been established, and the transportation facilities are supplied by two trunk lines of railroad and a navigable river. There are many hotels, besides several boarding houses, seven churches, eight newspapers (three daily and five weekly), two public halls, an opera house, a court-house, a high school, a driving park, fair grounds, etc., and also many wholesale houses, comprising dry goods, drugs, provisions, clothing, hardware, lumber and agricultural implements.

The banks of Fargo at present number four, two of which are organized under the National Banking Act and two as private banks. All own the buildings wherein they do business, the First National having a handsome two-story block on the corner of Front and Sixth Streets. The bank facilities are, however, scarcely sufficient for the volume of business, and must soon be increased. The Northern Pacific Railroad has here a round-house, repair shops and rail mills, each employing a large force. The Fargo Car Wheel and Iron Works and the Fargo Paper Mill Company are establishments which alone employ from 200 to 300 men. The several lumber yards annually sell many million feet of building and finishing material. Three planing mills, a brewery costing nearly $100,000, and a flouring mill, with a capacity of over 400 barrels of flour daily, give evidence of the progress which the city is making. The Northern Pacific Elevator Company has its head-quarters in Fargo, owning over fifty elevators and as many more warehouses scattered over Dakota and northern Minnesota on the lines of the Northern Pacific and Manitoba Railroads. Three large elevators, with a capacity of over half a million bushels, are in operation. The city has a prosperous building association, with $600,000 capital, which has already erected a large number of residences. The Fargo Improvement Company, with $200,000 capital, is also meeting to some extent the wants of new comers in this direction by the erection of business blocks and warehouses, as well as dwellings. The Chamber of Commerce, composed of the representative business men of the city, is an organization which exerts an active influence upon the best interests of the place. The principal streets and the larger business houses are lighted by electricity, and a tower, 200 feet in height, carries at its apex 20,000 candle-power lights. Fargo has a well organized and fully equipped fire department, with five companies. The educational facilities of the place are exceptionally good. The school-house, near the court-house, has been enlarged from time to time, as the constantly increasing needs of the population have demanded, and in 1880 a very fine brick building was erected, costing $8,000. The High School, situated on Adams Avenue, cost $40,000. Several ward schools are also to be established, blocks having already been set apart for this purpose in various parts of the city. The water supply is drawn from works constructed on the Holly system. The capacity of the works is 3,000,000 gallons per diem. Many miles of mains have been laid, and the requisite public hydrants are in place. The Northern Pacific Railroad has division head-quarters here, being the end of the Minnesota and the beginning of the Dakota Divisions, as well as the junction of the Fargo and Southwestern Branch. The city supplies farmers within a radius of at least fifty miles. *Canfield*, 7 miles west of Fargo, is simply a side track.

Mapleton—(12 miles west of Fargo; population, 450.)—This town possesses a steam elevator and warehouses, two hotels, one hall, a church, general stores, and one of the finest and best appearing school-houses in north Dakota. It is in the midst of a fertile region, and its prospects are bright, for it is peopled with energetic and enterprising business men.

Green—(15 miles from Fargo.)—This station is in the midst of the great bonanza farm, formerly known as the Williams farm, which is noted as having given its proprietors a profit of nearly $60,000 in the last two wheat crops, and affords better prospects each succeeding year. Mr. Green has the handsomest grove of young trees along the line of the Northern Pacific Railroad. Three miles west of Green is *Dalrymple Station*, the shipping point of a farm 20,000 acres in extent, which is owned by Mr. Oliver Dalrymple, the famous wheat-grower, after whom the place is named.

Bonanza Farming.—A peculiarity of wheat growing in Dakota is the grand scale upon which it is frequently conducted. Prior to 1875 it was declared upon high army authority, that beyond the Red River the country was not susceptible of cultivation; in going west from that stream to the James, there was some fair land, but much that was useless; and thence to the Missouri there was little or no available area, except the narrow valleys of the small streams; in fine, with the exceptions named, that the country was practically worthless. This sweeping statement gained wide publicity, and caused much hesitation with respect to undertaking the cultivation of the Dakota prairies. But Messrs. George W. Cass and Benjamin P. Cheney, both heavy capitalists and directors in the railroad company, having faith in the fertility of the land, determined to test its capacity for wheat production. They first bought, near the site of the present town of Casselton, 7,680 acres of land from the railroad company, and then secured the intervening Government sections with Indian scrip, thus obtaining compact farming grounds of enormous area. Mr. Oliver Dalrymple, an experienced wheat farmer, was engaged to manage the property, and in June, 1875, he turned his first furrow, plowing 1,280 acres, and harvested his first crop in 1876. The acreage was increased in each succeeding year, until in 1882 there were not less than 27,000 acres under cultivation. This immense farm does not lie in one body. One part of it, known as the Grandin farm, is situated in Traill County, thirty miles north of Casselton. The entire area embraced by the three tracts is 75,000 acres. Farming operations, conducted on so gigantic a scale, seem almost incredible to persons who are

Harvesting on a Bonanza Farm.
[By permission of Harper & Brothers, New York.]

only familiar with the methods of the older and more settled States. In managing the affairs of a "bonanza farm" the most rigorous system is employed, and the cost of cultivation averages about $1 per acre less than on smaller estates. The plan adopted by Mr. Dalrymple and all the other "bonanza" men, is to divide the land into tracts of 6,000 acres each, and these are subdivided into farms of 2,000 acres each. Over each 6,000 acres a superintendent is placed, with a bookkeeper, head-quarters building and a storehouse for supplies. Each subdivision of 2,000 acres is under the charge of a foreman, and is provided with its own set of buildings, comprising boarding houses for the hands, stables, a granary, a machinery hall and a blacksmith's shop, all connected with the superintendent's office by telephone. Supplies of every description are issued only upon requisition to the several divisions. Tools and machinery are bought by the car load from manufacturers; farm animals are procured at St. Louis and other principal markets; stores of every description, for feeding the army of laborers, are purchased at wholesale, and the result of the thorough system and intelligent economy in every department is found in the fact that wheat is raised and delivered at the railroad at a cost varying little from thirty-five cents per bushel. The net profit on a bushel of wheat is never less than forty cents, and the average yield per acre may safely be put at twenty bushels, although it often exceeds that quantity. Taking the lowest figures as a basis of calculation, the profits in 1882, on the 27,000 acres which Mr. Dalrymple had under cultivation, were not less than $216,000! No wonder that farming on this scale is called bonanza farming.

On this great farm, or rather, combination of farms—the 20,000 acre tract at Casselton—400 men are employed in harvesting, and 500 to 600 in threshing. Two hundred and fifty pairs of horses or mules are used, 200 gang plows, 115 self-binding reapers and twenty steam threshers. About the 1st of August the harvester is heard throughout the length and breadth of the land, and those who have witnessed the operation of securing the golden grain will never forget the scene. The sight of the immense wheat fields, stretching away farther than the eye can reach, in one unbroken sea with golden waves, is in itself a grand one. One writer describes the long procession of reaping machines as moving like batteries of artillery, formed en *échelon* against the thick set ranks of grain. Each machine is drawn by three mules or horses, and with each gang there is a superintendent, who rides along on horseback, and directs the operations of the drivers. There are also mounted repairers, who carry with them the tools for repairing any break or disarrangement of the machinery. When a machine fails to work, one of the repairers is instantly beside it, and, dismounting, remedies the defect in a trice, unless it prove to be serious. Thus the reaping goes on with the utmost order and the best effect. Traveling in line together, these 115 reaping machines would cut a swath one-fifth of a mile in width, and lay low twenty miles of grain in a swath of that great size in the course of a single day. "Carleton," a correspondent of the *Chicago Tribune*, described the reaping scene thus:

"Just think of a sea of wheat containing twenty square miles—13,000 acres—rich, ripe, golden—the winds rippling over it. As far as the eye can see there is the same golden russet hue. Far away on the horizon you behold an army sweeping along in grand procession. Riding on to meet it, you see a major-general on horseback—the superintendent; two brigadiers on horseback—repairers. No swords flash in the sunlight, but their weapons are monkey-wrenches and hammers. No brass band, no drum-beat or shrill note of the fife; but the army moves on—a solid phalanx of twenty-four self-binding reapers—to the music of its own machinery. At one sweep, in a twinkling, a swath of 192 feet has been cut and bound—the reapers tossing the bundles almost disdainfully into the air—each binder doing the work of six men."

Casselton—(20 miles west of Fargo; population, 1,300)—is a thriving, bustling town, the situation of which is very advantageous, being in the midst of one of the finest wheat-raising districts in Dakota. The first house at Casselton was built by the railroad company in 1877, and during that winter there were only four inhabitants in the place. In the spring of 1878 the first business house was put up, and during that season several others were erected, and some residences. Improvements have been going on ever since, and the growth has been steady. The town has no bonded indebtedness, which speaks well for the business qualifications and thrift of the community. It was incorporated in the summer of 1880, the governing board holding its first meeting on August 4th. The place has an organized fire department. Sidewalks have been put down, and a liberal appropriation is annually made for grading and improving the streets. The business of Casselton is represented by five houses which sell general merchandise, one grocery store, two drug stores, two harness shops, two meat markets, four agricultural machinery stores, eight livery and sale stables, five hotels—one with accommodations for 200 guests—two blacksmith shops, one bank (the First National), one lumber yard and one wheelwright shop. There are three elevators, with a capacity of 200,000 bushels, and a large and well equipped flouring mill, as well as a brick yard, manufacturing a superior quality of brick, one public hall, a newspaper and four churches. The public schools are efficiently organized under the graded system, and are in successful operation. There has been constructed already one commodious ward building, and in addition to this, there is a high school building erected at a cost of $12,000. The Presbyterian denomination of Dakota have established their educational institution at Casselton, and named it "Casselton University." This institution is fully incorporated, and the board of trustees is actively at work pushing the enterprise forward. The denomination has pledged to its endowment $10,000 during the next three years, and the citizens of Casselton $20,000. The farmers of Casselton, in speaking of the excellence of their opportunities, say that they do not suffer materially, either in wet or dry seasons. The farms lie just high enough to be secure from the overflow of the Red River. Good well water can be obtained at a depth of twenty-two to twenty-five feet. There are three artesian wells in the vicinity, one being six and a half miles south of this point, another nine miles north, and the third at the Casselton mill.

From Casselton a branch line of the St. Paul, Minneapolis and Manitoba Railroad runs northwest, and the Breckenridge branch of the same railroad crosses the Northern Pacific three miles west of the town.

Wheatland —(27 miles west of Fargo; population, 600.)—This town is established upon the dividing ridge that separates the magnificent black soil of the Red River Valley from the undulating prairie beyond toward the Cheyenne, and is supplied with general stores, farmers' supply depots, two hotels, a school-house, which is also used for church purposes, a newspaper and an elevator of 60,000 bushels capacity. It is the trading point for numerous small farmers, and also the head-quarters for several large bonanza farm interests in the vicinity.

New Buffalo—(37 miles west of Fargo; population, 200.)—New Buffalo is an incorporated village, and the trading point for farmers in its vicinity, the exports being principally wheat, oats and potatoes. It has an altitude of 575 feet above the level of Fargo. The surrounding country is an even, unbroken prairie, as far as the eye can reach. The first settler came to New Buffalo in 1878, and took a claim about one-half mile north of the present village. The town was laid out in May, 1878, and the first house was occupied as a store, post-office and dwelling. The same year the railroad depot and a blacksmith shop were erected. There are an elevator, with a capacity of 75,000 bushels, a newspaper, four general stores, representing various branches of trade, three hotels and a harness shop. North of New Buffalo, and adjacent thereto, will be found the bonanza farms of ex-General Manager Sargent, of the Northern Pacific; Col. Rich, of Michigan, T. D. Platt and others, all of which produce large crops of wheat.

Tower City—(42 miles west of Fargo; population, 800.)—This town, named in honor of a former director of the Northern Pacific Railroad, is on the western edge of Cass County. It was laid out in April, 1879, when there was no settlement nearer than Valley City, sixteen miles westward. The growth of the town has not been rapid, but it has been, nevertheless, steady and healthy. The population is chiefly made up of Americans, Canadians, Germans and a few Scandinavians. The soil of the surrounding country is the rich, dark vegetable loam which characterizes Cass County. Tower City has three church organizations, Baptist, Presbyterian and Methodist, with substantial buildings; a school-house costing $1,500, besides two hotels, bank buildings, substantial business blocks, handsome residences, a public hall, a newspaper, a steam elevator and a bed spring manufactory. The Tower City Milling Co., with a capital of $20,000, has in course of construction a steam flouring mill, the capacity of which will be 125 barrels per day.

An Artesian Well.—The Northern Pacific Railroad, in boring a well at Tower City, struck a vein of water at a depth of 670 feet. The water is soft, not very cold, sweet and pleasant to the taste, and its medicinal properties are said to be similar to those of the springs at Saratoga. Many persons who use the water say that it works on the kidneys in a beneficial manner and tones up the entire system. The city boasts of two pretty parks—the Ellsbury, which is situated on the north side of the railroad, bordering on Michigan Avenue, and the Villard, just south of the railroad depot. The latter is enclosed by a substantial fence, and the trees which it contains are thriving satisfactorily. In the centre of the park is a fountain, supplied with water from the artesian well. On the arrival of a train the travelers usually make a rush for the fountain, for the purpose of testing the medicinal water.

Oriska—(48 miles west of Fargo; population, 200.)—This place, situated midway between Fargo and Jamestown, is surrounded by thousands of acres of fertile prairie, dotted with many lakes of pure water, and a more desirable farming and stock country could scarcely be found. The soil is of the first grade and of great depth, with a clay subsoil. Good water is abundant at a depth of from ten to twenty feet, being entirely free from alkaline salts and as clear as crystal. The nutritious grasses for which this country is noted, and upon which stock thrive so finely, attain a luxuriant growth, large quantities of hay for wintering stock being cut each season. Oriska has an hotel, a church, a school, general stores in all branches of trade and an elevator with twenty thousand bushels capacity.

Valley City—(58 miles west of Fargo; population, 1,500)—is the county-seat of Barnes County. It lies in a deep valley surrounded by an amphi-theatre of hills, which rise to a height of 125 feet or more on every side of it. Circling around the valley is the beautiful Cheyenne River, a stream at this point fully seventy-five feet in width, running over gravelly beds, and fringed with sturdy oaks, elms and other woods. The Northern Pacific Railroad enters the town on its eastern side by a winding passage through the bluffs for a distance of several miles, and emerges on the steepest part of the line between Fargo and the Missouri River. The town is furnished with a fine water power by a fall of ten feet in the river within the limits of the city proper. It was evident from the day that the railroad first made the place accessible, that a city would spring up at this point which would become one of the most important in northern Dakota. The Cheyenne River, to which the town owes much of its prosperity, is one of the few important rivers of Dakota. It rises in the northern part of the Territory, in the vicinity of Devil's Lake, and describes a tortuous course of nearly 100 miles before it reaches Valley City. Its waters are generally clear and abound with fish, and its banks are skirted with timber. Along its shores in former years roamed the savage Sioux, and many a bloody conflict has taken place between warrior tribes within sight of its wooded slopes. Twenty years have passed away since the Indians were driven across the Missouri, and the only mementoes of the red men to be found to-day are the bones of the buffalo, which lie bleaching everywhere over these Dakota prairies. The town is finely provided with wide streets and avenues, and the business and residence structures are of attractive appearance. A large proportion of the population are Americans, but there are also many pros-perous, enterprising Scandinavians and Germans. Among the public buildings are an imposing court-house which cost $35,000, having ample accommodations, not only for the county officers, but for the United States Court; a brick jail built at an expense of $10,000; one large brick hotel costing over $30,000, besides two smaller frame ones; an opera house, sev-eral churches and school-houses, three National banks built of brick—the edifice occupied by the First National having been erected at a cost of

$15,000. There are three good newspapers. The city has two brick yards, of which the output is about 20,000 bricks of fine quality a day. It is also well supplied with lumber and coal yards, and has two large flour mills operated by power from the Cheyenne River. During the year 1882 the improvements showed an outlay of nearly $215,000, and from present indications the building operations in 1883 will no doubt be forty per cent. larger. The receipts and shipments of wheat show that the growth of the city is not in advance of the country surrouning it, but of a most healthful kind. The business of the elevator in 1882 was 268,000 bushels, or an increase of 164,000 bushels over the preceding year, not taking into the account the many thousands of bushels received and ground at the Cheyenne roller mills, and leaving out entirely any mention of the unthreshed and unmarketed grain, estimated by dealers to be fully 175,000 bushels. *Hobart*, eight miles westward, is a small station, with an elevator for handling wheat.

Sanborn—(71 miles west of Fargo; population, 700.)—In 1880 there was scarcely any population in the neighborhood of Sanborn, but now there is a good town here, with a great deal of land occupied and cultivated. This is shown by the fact that the Sanborn elevator received about 100,000 bushels of grain in 1882. The town has been stimulated to a rapid growth mainly by the energy of a single enterprising business firm, and is fast acquiring an influential position. Its situation, in the midst of rich fields, with good farms a short distance away depending upon Sanborn for supplies, is very advantageous. In 1882 the average yield of wheat was about twenty-five bushels of "No. 1 hard," with the ground only half prepared at that. Sanborn has two newspapers, a public hall, two hotels, two churches, two elevators, a good school, two banks, several large business establishments, including hardware and agricultural implement stores, harness shops, etc., all of which are well patronized by the farming community. The products are wheat, oats and barley. A few antelope, ducks, grouse and geese may be found in the vicinity. The Sanborn, Cooperstown and Turtle Mountain Railroad is under construction and is running north some twenty-five miles.

Eckelson—(74 miles west of Fargo)—is a new town, situated on Lake Eckelson, a lovely sheet of water. The land is high and rolling, the soil as rich as any in the region, and by virtue of the lake, which is thirty feet below the level of the town, a natural and perfect system of drainage is provided. Lake Eckelson—seven miles long and three-quarters of a mile wide—affords excellent opportunities for bathing, fishing and boating. A colony from Elmira, New York, has permanently established itself at this point, its members having laid plans for building improvements and farming operations on a large scale. Aside from several neat dwellings, there is a substantial and commodious school-house, which was constructed at a cost of $3,000, and which speaks well for the people who have made this their permanent home, two general stores, an elevator, depot, hotel, and other business establishments.

Spiritwood—(83 miles west of Fargo; population, 100.)—Spiritwood is in the midst of a fine grain-growing country, and has several bonanza farms around it, making the town an important shipping point. The village contains one store, one school, and an elevator with 50,000 bushels capacity. Spiritwood Lake is a very beautiful spot, and its waters teem with pickerel, bass, perch and some smaller kinds of fish. Ducks, geese and prairie chickens also abound in this vicinity.

Jamestown—(94 miles west of Fargo; population, 2,500.)—The town is the county-seat of Stutsman County, 368 miles west of St. Paul, and is excellently situated in a rich agricultural region on the east bank of James River, surrounded by ranges of handsomely sloping hills. It is a bright, new place and growing rapidly in importance. The town was incorporated in June, 1880, with 400 inhabitants. The streets are wide and regularly laid out, with good sidewalks, and the character of the buildings is substantial. There are three banks, with an average deposit of $300,000; six hotels, three churches, a large flouring and planing mill, the former costing $15,000; excellent public schools, among which is one erected at a cost of $15,000, with the names of 300 pupils upon its roll; two elevators, two well conducted newspapers, brick and lime kilns, several large and well stocked mercantile establishments. Coal of good quality for household purposes is accessible and cheap; the climate is healthful and invigorating, and the citizens are enterprising and intelligent. In fact, all the elements of a rapid and permanent development exist at Jamestown. In 1882 between 150 and 200 buildings were erected, at a cost of over $600,000, and there were also heavy expenditures on sidewalks and other improvements. Among the most important of the new structures were a $35,000 court-house, a $20,000 elevator, two bank buildings at $25,000 each, and the Northern Pacific round-house and machine shops at a cost of $75,000. The report of the Jamestown Board of Trade furnished the following statistics for the year 1882: The roller flouring mill, with a capacity of 100 barrels of flour per day of twenty-four hours, made 14,522 barrels of fine flour, sold of bran and shorts 1,060 tons, and handled some 75,000 bushels of No. 1 hard wheat. The brickmakers reported the manufacture of 2,300,000 bricks, at a cost of over $16,000, sixty men being employed in the work. The approximate number of bushels of grain raised in the county in 1882 was 600,000, and double the amount of acreage was prepared for the crops of 1883. The hay cut of 1882 probably footed up 2,500 tons, and sold at $10 per ton. The elevators and mill reported that 300,000 bushels of wheat, 30,000 bushels of oats and 13,000 bushels of barley were received during the year 1882.

Jamestown is surrounded by a rich agricultural district, and from its natural situation is becoming one of the most important commercial and railroad centres in the new Northwest. The whole of the upper James River Valley is practically tributary to the city, as are also the Devil's Lake and Mouse River countries. This last-named region, which is very favorably looked upon, has been opened up to settlement by the Jamestown and Northern Railroad, of which Jamestown is the southern terminus.

Stutsman County is 36 by 40 miles in extent and covers an area of 1,105,-920 acres, nineteen-twentieths of which are available for cultivation and pasturage. There are four streams in the county, three of which are timbered along their banks with hard wood. The banks of the streams are well defined, and contain deposits of granite and limestone, which are excellent for building purposes. There are also several fine lakes in the county fringed with trees, and the lakes and streams abound with fish. Stock raising has been thoroughly tested as a branch of industry and is a recognized success. Cattle, horses, sheep and hogs alike thrive.

North of Jamestown can be found the "Hawksnest," where Gen. Sibley had the Sioux corraled at one time. There are several battle fields in the vicinity where fierce conflicts took place between the troops and the Sioux.

The Coteaux of the Missouri, situated twelve to twenty miles away, furnish antelope and an occasional buffalo. Stages run daily north to Ft. Totten and Devil's Lake, and tri-weekly south to Tarbels, eight miles; Ypsilanti, fourteen miles; Montpelier, twenty miles, and Grand Rapids, thirty-five miles.

Jamestown and Northern Branch.

JAMESTOWN TO CARRINGTON.—DISTANCE, 43 MILES.

This branch line has partially opened to settlement the extensive agricultural lands of the Souris or Mouse River, and the Devil's Lake or Minnewaukan region. Five farming towns sprung into existence along the railroad before it was finished to Carrington, and settlement has since pushed ahead in advance of the track. The names of the stations are *Arctic, Buchanan, Pingree, Melville* and *Carrington*, respectively distant from Jamestown six, thirteen, twenty-one, thirty-four and forty-three miles. Each of these places will develop in the course of a few months into a lively centre of trade, furnishing the neighboring farmers with supplies and shipping their crops.

Carrington, the county-seat of Wells County, is already a prosperous town, established in the spring of 1883 by the Carrington and Casey Land Company, a corporation owning considerable tracts of land in the Upper James River Valley. The place has had a very rapid growth, and probably now contains about 500 inhabitants. It is in the centre of one of the finest wheat-raising districts of northern Dakota. The road is finished beyond Carrington to the crossing of the Cheyenne River, a distance of about forty miles, and will soon be opened to the western end of Devil's Lake, where a connection will be made with steam-boat navigation on the Lake.

A branch of the Northern Pacific Railroad system, twenty miles in length, has been built from Carrington to *Sykeston*, a place of some importance. This branch will probably be extended to the bend of the Mouse River. Sykeston derives its name from an enterprising English gentleman, Mr. Sykes, who purchased a large area of excellent farming land lying west of Carrington, near the base of the Coteaux.

The country which is penetrated by the Jamestown and Northern Branch Railroad, in its general features, presents the aspect of a broad prairie, bounded at the west by the Coteaux of the Missouri, and extending eastward beyond Devil's Lake, cut by the narrow valleys of the James, Cheyenne and Mouse rivers, and relieved in the monotony of its landscape by low hills abruptly rising here and there. The rivers, as well as most of the small lakes so frequently met, are fringed with a good growth of oak, ash, elm and black alder, and the land is covered with stout, rich meadow grass. Along the big bend of the Mouse River many seams of lignite coal crop out of the bluffs, some of which are of sufficiently good quality to be mined. Devil's Lake is about fifty miles in length and ten in width. Its water is salt, but many kinds of fish thrive in it. Most of its southern shore is occupied by a reservation of the Sisseton band of Sioux Indians, who are peaceable and tolerably industrious, having cattle and cultivating small fields. A belt of valuable timber skirts nearly the entire shore line on the northern side, and the alluvial prairies stretch out in every direction, affording a fine opportunity for engaging in stock raising and general farming. This inviting region is attracting large numbers of settlers.

Dakota Division—Main Line.

[*Continued from page* 104.]

Eldridge—(101 miles west of Fargo; population, 100.)—This village, which has only been settled a short time, contains a store, a church and school building combined, a hotel and an elevator of 10,000 bushels capacity. The products are wheat, oats, barley and potatoes. *Thackeray*, seven miles beyond, is simply a side track.

Cleveland—(112 miles west of Fargo; population, 100)—is one of the prettiest town sites between Jamestown and Bismarck. Surrounded by good agricultural and grazing lands, the place is settling up quickly. Farming was begun in 1883 upon the heretofore uninhabited prairie. The soil is eighteen to twenty-four inches deep, with a clay subsoil of eighteen feet. The town contains a depot, side track, post-office, two stores, lumber yard and telegraph office. *Medina*, nine miles further west, is at present an unimportant station, although it is much resorted to by sportsmen, who find good hunting and shooting in the vicinity.

Crystal Springs—(130 miles west of Fargo; population, 100.)—This is a new town, with excellent outlying agricultural lands, and good crops of wheat, corn, oats and potatoes are produced. The small lakes not far distant contain quantities of fish. Deer, elk, antelope and a variety of small and feathered game abound in the neighborhood. The village has a general store, a convenient hotel, and will doubtless develop into a centre of trade.

Tappan—(139 miles west of Fargo; population, 115.)—This is one of the handsomest small stations on the line of the railroad. It is situated on the Troy Farm, where about 2,500 acres are under cultivation. This farm was established in 1879, receiving its name in compliment to two of its owners, the station and post-office being called Tappan in honor of a third proprietor. The farm consists of sixteen sections, or 10,240 acres, embracing most of the railroad land in two townships. In 1882 over 2,000 acres were cultivated, and 400 acres of new land were broken. Mr. Van Deusen, the manager of this extensive farm, recently wrote that in 1880 "we cropped 1,300 acres, the wheat running twenty-five bushels per acre, oats, fifty bushels, and barley, forty bushels. In 1881 the crop was as near a failure as it is possible to conceive in this country. The wheat that year averaged over ten bushels, but the price brought us out all right, as we sold at from $1.35 to $1.50. The crop of 1882 was a good one, and for my own satisfaction I had the entire product of 600 acres of wheat weighed. This lot of land produced a little over 16,000 bushels of No. 1 Hard. Had small portions of this body been taken separately, the yield would have shown much heavier. The balance of the land did fully as well. In common oats 150 acres averaged fifty-eight bushels, weighing forty pounds to the bushel. The barley, 400 acres, ran thirty bushels, as handsome and heavy as you ever saw. The white Russian oats went sixty-eight bushels per acre. We put in a little winter rye last fall for spring feed for the mules. It was very late coming up, and we were concluding to plow it up when the stalks appeared. The story seems large, but it is a fact that the two and a half acres of this grain threshed out seventy-nine bushels, an average of thirty-one and two-fifths bushels per acre. We have a lot of white beans, which are handsome and very heavy. We can raise any kind of root crops in proportion. The kitchen garden is filled with all sorts of vegetables, from the egg plant down. We find the size of our melons and melon crop depends largely on the size of the crew of men on the farm. We run fifty-six horses and mules, and their perfect health is remarkable. We have not lost an animal since we have been here. It is certain that this will eventually be one of the great stock countries of the States. We have kept sheep for three years, the Cotswold, and have not lost a head

We have only 100 head, but they keep in prime condition, as do our cattle, on prairie grass, hay and straw. None of them have ever had a particle of grain."

Dawson—(144 miles west of Fargo; population, 350)—is an enterprising town near the centre of Kidder County. It is growing rapidly and is an important business point. An excellent agricultural and stock-raising country is tributary to it. It has three hotels, one of which cost $15,000, six general stores, lumber yards, livery stables and a newspaper. The products are wheat, oats, barley and potatoes. About two miles south of the depot lies a beautiful body of fresh water, called Lake Isabel. J. Dawson Thompson, from Pennsylvania, who has a farm at Dawson, wrote in the autumn of 1882:

"I came here two years ago for a trip, and fell in love with the country. I exercised all my rights with the Government, and secured besides enough land to aggregate 3,000 acres. Last year I broke 400 acres. This fall I harvested 300 acres of wheat, No. 1 hard, running thirty-two bushels to the acre, and 100 acres of oats running seventy bushels to the acre. The wheat thus far has sold at ninety-two cents net. The oats are worth fifty cents. I also had ten acres of potatoes, that yielded 350 bushels per acre. I raised a full line of vegetables that yielded as well as any I ever saw. I put in three acres of corn for a trial. It matured well and ran seventy bushels per acre. The crop on the 400 acres, in 1882, yielded more than twice the original cost of my land."

Steele—(151 miles west of Fargo; population, 350)—is a thriving town near the western boundary of Kidder County, of which it is the county-seat. The place is a monument of the energy and perseverance of its founder, Mr. William F. Steele, who has within three years transformed the prairie, 1,300 feet above the level of the sea, and the highest point on the line of the Northern Pacific between Duluth and Bismarck, into a town with a fast growing population. Situated in a rich agricultural district, Steele is already a favorable trading point, containing general stores, a court-house, two hotels, a newspaper, an elevator and lumber yards, a school and a church. Mr. Steele has at least 3,000 acres under cultivation, and has been uniformly successful with his crops, his wheat yielding above the average, in consequence of his thorough and systematic farming. There are many small farms in the neighborhood of Steele, the owners of which are quite prosperous. Four miles south of the town is a beautiful lake, full of fish, and the hunting opportunities are also excellent.

Geneva, Driscoll, Sterling, McKenzie, Menoken and *Apple Creek*, situated on the line of the railroad, distant from Fargo, respectively, 158, 164, 170, 176, 182 and 191 miles, are at present shipping and supply stations in the midst of a fine agricultural and grazing region. They are likely, however, under the impetus of enterprising settlers and business men, who are cultivating lands and raising stock in the neighboring country, to develop into thrifty villages. At each of these points there is one or more general supply stores, and the population and business interests are constantly increasing. Sterling had about 100 inhabitants in the spring of 1883, and Menoken has been for a year or two the distributing and shipping point for the large farms which surround it. Apple Creek is a beautiful stream, offering fine sites along its banks for farm houses.

Tree Planting.—It has been fully demonstrated for a long time past that many forest trees will grow and thrive under cultivation upon the naked prairies and plains of the Western States. The advantage of timber to the settler admits of no dispute. To encourage and stimulate the farmers of the Dakota plains to engage in tree planting has been one of the aims of the railroad company. To accomplish its purpose most effectively, it has itself organized a tree planting department, and made a liberal appropriation to cover the expense of the work for a period of five years. Active operations were begun in the spring of 1882 at Tower City, Tappan and Steele. These points have distinctly defined and different characteristics of soil, and so are well suited to testing several varieties of trees. At Tappan 200,000 trees and cuttings were planted, the same number at Steele, and also ten bushels of box elder seed, from which have sprouted about 300,000 thrifty shoots. At Tower City there were likewise planted 100,000 trees and cuttings. Only a small percentage of these trees have died, and the mass are in a vigorous condition. The varieties thus far planted are white willow, cottonwood and box elder, but the experience which has been gained, and additional knowledge of the soil and climate, justify the opinion that the list of forest trees available for planting can be enlarged, so as to include the white maple and ash. The company has also broken about 600 different patches of ground along 200 cuts on the line of the railroad between Fargo and Bismarck, upon which trees will be planted to serve as wind breaks. It has also offered a series of premiums to farmers to plant groves and ornamental trees about their premises. This fact, in connection with the liberal provisions of the Tree Culture Act of the United States Congress, will doubtless insure, in the course of a few years, a growth of forest and ornamental trees at intervals on the open plains of Dakota.

Bismarck—(195 miles from Fargo; population, 3,500.)—This is the county-seat of Burleigh County, and on the 2d of June, 1883, was chosen as the capital of Dakota Territory. The geographical position of Bismarck is scarcely inferior to that of any city between the Atlantic and Pacific oceans. It is situated on the east bank of the Missouri River, which, with its tributaries, gives about 2,000 miles of navigable water above it to the northward and westward, and the same number of miles below it southeastward to St. Louis. Its landing is one of the finest on the great river, and the place has already become, and is likely always to remain, the centre of steam-boat navigation in the Northwest.

The town is as remarkable for its healthy situation as it is for the productiveness of the land which environs it. Its elevation above the sea is 1,690 feet, and it not only lies above the line of possible submergence by the river, but is well adapted to easy and cheap drainage. Not more than ten years ago was the project of building up a city at this point first entertained. In 1872 the engineers of the Northern Pacific Railroad decided upon crossing the Missouri here; and this decision, supplemented by the local and surrounding advantages, resulted in the survey and first settlement of the city of "Edwinton," soon after changed to Bismarck by resolution of the Board of Directors of the railroad company.

Bismarck is one of the oldest cities in Dakota, and has up to the present time experienced a somewhat slow though substantial growth. Recently the city has begun to develop very fast, and it is already one of the most prominent towns on the line of the railroad. Substantial county buildings, banks, the *Tribune*, a well conducted daily and weekly newspaper, good hotels, public halls, a high school and other educational establishments, four churches, large business blocks, fine residences, flouring and lumber mills and an elevator demonstrate the prosperity of the place. An artesian well is sunk on a hill overlooking the city, where a reservoir may be easily made, and it is believed that sufficient water will be obtained to meet future needs.

Camp Hancock, named after General Hancock, is the United States Army Depot and Chief Quartermaster's Head-quarters. The Quartermaster keeps the ground in nice order, has a good house, neat office and extensive stabling. The supplies for Forts Lincoln, Yates, Stevenson, Buford and Assiniboine come here and are reshipped by river, this being an important shipping point for military supplies and Indian goods. It is also a port of entry, goods in bond going through Bismarck, billed for the Northwestern British Possessions, 1,500 miles distant. At the steam-boat landing there are several warehouses through which the goods for the river route pass. Twenty-five different steam-boats arrive and depart from this landing, carrying on an important commerce. The "store trade" of one line of four boats amounted to $125,000 in a single season. This trade means that the steam-boats furnish supplies to sutlers, wood choppers and others who occupy the river country at different points between Bismarck and Fort Benton, Mont., the head of navigation. The return freights are wool, buffalo robes and hides, and bullion.

Surrounding Bismarck are wide expanses of arable soil, which are capable of producing everything necessary to the subsistence of a large population. Ex-President R. B. Hayes has a large farm five miles north of the city, from which he takes heavy crops of wheat, and where he is raising thoroughbred animals. Land is taken up on the east side of the Missouri as far north as Fort Stevenson, and at Painted Woods, about twenty miles north, are thriving settlements of Scandinavians, Russians and not a few Americans. Several of the farmers have large herds of cattle and sheep, but wheat and oats are the staples.

Bismarck is an excellent starting point for parties in pursuit of antelope, deer, elk, bear and feathered game, as outfits can be purchased at fair prices and good guides secured for the country north of it.

Fort Abraham Lincoln.—An eminence of easy ascent, within the city limits of Bismarck, has received the name of Capitol Hill, and its summit is soon to be crowned with state buildings of imposing architecture. From this point a wide and beautiful view is obtained—a prominent object in the scene being the white walls of Fort Abraham Lincoln. This military station lies five miles distant by the road, on the high bluffs on the west side of the Missouri, and not far from Mandan. It was attacked on five different occasions, during the years 1872-73, by the Sioux, with an aggregate loss of eight killed and twelve wounded on the side of the troops, but the repulsed Indians suffered more severely. The gallant and ill-fated General Geo. A. Custer passed the last two years of his life at this post. One of the friends of the deceased general, in describing the incidents of Custer's busy life, gives a glimpse of his room at the old fort in the following words: "It was pervaded by an air of luxury and good taste, although the furniture was of the plainest, and much of it old and worn. But over every old chair or sofa, covering all deficiencies, were beautiful furs and skins that money could hardly purchase—the spoils of Custer's rifle, and all around the walls hung grand heads of buffalo, of 'bighorn,' of graceful antelope—heads prepared by Custer himself; the fierce faces of wolf, bear or panther giving a wild and peculiar grace to the lofty room lit up by the glow from yonder ample fire-place, with its blazing logs."

The Great Bridge Over the Missouri River.—This superb bridge was opened for traffic with appropriate ceremonies on the 1st of October, 1882. Prior to that time the river was crossed by means of a large transfer steamer, specially constructed for the purpose of carrying trains of cars. Owing to the strong current and constantly shifting sand bars in the channel, the ferriage by steam-boat was always tedious, and the Northern Pacific Railroad Company never intended that so slow a transfer of its trains should be anything but a temporary arrangement. The Board of Directors, therefore, took definite action with reference to the construction of the bridge during the winter of 1880, and in the autumn of the same year the point was fixed at which the crossing should be made. The building of the bridge was placed in the hands of Mr. George S. Morison, an eminent engineer, under whose supervision every step in the prosecution of the great work was taken, the result of his directing skill being a structure as substantial as it is graceful. The cost of the bridge was about $1,000,000, and this large expenditure was ungrudgingly made by the railroad company, in view of the permanent character and engineering excellence of the work.

The Missouri here has the same marked characteristics that exist for its whole length. It is a silt-bearing river of the first magnitude. Deep below the surface lies a hard, stratified material—sometimes indurated clay, sometimes rock—but above this its bed is simply a mass of moving silt, which it has itself brought down from the disintegrating soil of the mountains, over which the river seems to maintain supreme control, working to the right or left or downwards. At the site where the Bismarck bridge now stands, during the progress of the work the depth of water varied from nothing to fifty feet, with a change in the surface elevation of not over ten feet, the flood washing out in its earlier stages the bottom of the river, which the same flood replaced with material brought from some point farther up, within a few days thereafter. A descriptive outline of the bridge and the methods which were pursued in its construction is here appropriate.

The Missouri River, at this point, is 2,800 feet wide, with a variable channel, about two-thirds of the whole width of the river, which is occupied, except at extreme high water, by sand bars. It was necessary to build a dike from the west shore to within 1,000 feet of the east shore, which is here a high bluff of extremely hard clay, in order to confine the river within a width favorable to the maintenance of a fixed channel. The construction of the dike was begun in the autumn of 1880, but, owing to the shifting character of the channel, the state of affairs was at first perplexing. Before any work was actually begun on the dike the channel had passed to the west side of the river, and the winter of 1880-81 found the work in very imperfect condition. It therefore became necessary to make a strong effort to prevent the destruction, by the breaking up of the ice, of the work which had already been done. On the 30th of March the ice moved out with unusual violence, the river rising thirteen feet above the ordinary summer floods, overflowing the entire bottom lands; but the ice passed, leaving the dike comparatively uninjured. The channel, however, remained near the west shore.

Not until April, 1881, did the construction of the bridge fairly begin. This involved three totally different pieces of work, *viz.*, the control and rectification of the river, the bridge proper, and the approaches. The control and rectification of the river consisted in confining its channel to 1,000 feet between the east shore and the end of the dike, and the protection of the east shore with rip-rap to render it secure from the eroding action of the water. After the disappearance of the ice, the main channel was left between the unfinished dike and the west shore, and the action of the current during the spring and summer floods wore away about 200 feet of the dike. The relative amount of water passing through the east and west channels changed gradually, however, until, about the end of April, 1881, the volume passing through the east channel was decidedly more than that through the west. In May of the same year the attempt to close the west channel by driving piles for a bridge to connect the west end of the dike with the west shore was defeated by high water before the work could be finished. A second attempt was made in July, after the summer floods had subsided, to perform this work, at which time it proved successful—the track then laid to the end of the dike having been maintained ever since. During the spring and summer of 1881 the east end of the dike was strengthened by rip-rap, which was transported by boats, until the before mentioned track was completed, when the material was unloaded directly into position from cars. By these means the end of the dike was maintained for many months against the strongest current in the river, and with a depth of fifty feet of water immediately outside of it. The west channel silted up rapidly during the fall months, and in the early winter of 1881–1882 was finally closed by filling the pile bridge with earth. This dike has since admirably maintained itself against the full current of the river, and is expected to do so forever. There was used in its construction 30,000 tons of granite boulders for rip-rap, besides a large quantity of brush and crib logs, and upwards of 20,000 cubic yards of clay. By means of the dike the river has been permanently confined to a width of 1,000 feet, adjoining the east shore. The course of the channel is gradually improving, and the indications are that it will soon follow an easy curve along the east bluff, leaving the dike buried in a deposit of silt. The elevation of the top of the dike was purposely fixed at about the level of the ordinary summer floods, in order to secure a larger deposit, both above and below the dike, than if it had been finished at a higher level. When the ice goes out, about the 1st of April, the water usually rises some feet above the top of this dike, and a secondary embankment, intended only to resist the action of the ice, has been built immediately alongside of the trestle approach, nearly twenty feet higher than the top of the dike.

The bridge proper consists of three through spans, each measuring 400 feet between centres of end pins, and two approach spans, each 113 feet. It is a high bridge, the bottom chord of the three main spans being placed fifty feet above the level of the highest summer flood, thus giving room for steam-boats to pass at all navigable stages of the river, the bridge allowing practically four feet more room than many of the bridges on the lower Missouri. The variable channel and the high bluff on the east side, were sufficient reasons for adopting the high bridge plan in preference to the low bridge with a draw, and the violent action of the ice added to the force of these reasons. The east end of the east approach span is supported by a small abutment of granite masonry founded on the natural ground of the bluff. The west end of the west approach span is upheld by an iron bent, resting on two Cushing cylinders, which are supported by piles driven into the sand bar. The three long spans are supported on four granite piers, which are of unusual size, with long raking ice breakers, shod with steel. They are fashioned so as to cut readily the large sheets of ice upon the breaking up of the river in the spring, and to afford the least possible obstruction to the moving mass of broken ice which follows. Their stability far exceeds any force which the ice can exert.

Pier 1, the easterly pier, rests on a concrete foundation, the base of which is twenty feet below ordinary low water, and sixteen feet below the estimated extreme low water due to ice gorges. Piers 2 and 3, which are in the channel of the river, are founded on pneumatic caissons, sunk into the underlying clay to a depth of fifty feet below ordinary low water and ten feet below the surface of the clay. Pier 4 is situated on the sand bar on the west side of the river, below the protection of the dike, and rests on a foundation of 160 piles, which were driven by a steam hammer.

Ground was first broken at the bridge site on May 12th, 1881, when the small excavation was made for the east abutment. On the 15th of July, 1881, the excavation for the foundation of Pier 1 was begun in earnest, and carried down through hard stratified clay, similar to that which underlies the whole of this part of Dakota, and on the 1st of October the concrete filling of this foundation was finished. The laying of masonry was begun on the 4th of October, and the pier was ready on November 28th, 1881.

The caissons on which Piers 2 and 3 are founded are built of pine timber, sheathed with two thicknesses of three-inch oak plank. They measure seventy-four feet long by twenty-six feet wide, and are seventeen feet high on the outside. Each caisson contains 133,000 feet of timber and 82,000 pounds of iron, besides nearly 500 cubic yards of concrete.

These caissons were built on shore, launched and then towed into position. The lower portion of each caisson formed a working chamber seven feet high, with flat roof and inclined sides. The upper portion of each caisson was a crib work of timber, filled throughout with Portland cement concrete. After the caisson had been towed to its place, the concrete above the working chamber was put in, the air-locks placed in position and air pumped into the working chamber, which was thus converted into a great diving bell. A force of men were put to work in this chamber to excavate the sand, which was carried off by columns of water, the caisson gradually sinking by its own weight as the excavation proceeded. The masonry was laid on the roof of the caisson and continued as the sinking progressed, the top of the masonry being always kept above water. This laying of masonry on a sinking foundation was a source of serious perplexity to the masons, who were greatly troubled when they found they could no longer make use of a level to set their stone.

The caisson for Pier 2 was launched August 6th, 1881. A sand bar had formed in front of the launching ways, and it was necessary to drop the caisson a quarter of a mile down stream and tow it up against the current. This towing was done by the transfer steamer, "Northern Pacific, No. 1," and was probably the hardest service which this craft was ever called upon to perform, towing a large and heavy caisson drawing five feet of water having taxed her capacity to the utmost. On the 12th of August the caisson was correctly placed in position, and the air pressure was soon afterward put on. On the 1st of September the laying of masonry was begun, and on the 29th the caisson grounded on the clay. Early in November the process of filling the working chamber was begun, and on the 16th of that month the entire working chamber, air lock and shafts having been filled with concrete, the foundation was completed. On the 19th January, 1882, the masonry of this pier was done.

The caisson for Pier 3 was launched October 21st, 1881, and the concrete filling begun on the 26th. Air pressure was applied on November 24th, using the machinery which had been relieved from work at Pier 2. On the 16th of December clay was struck, and on the 17th January, 1882, the foundation was finished. The laying of masonry was interrupted by the spring flood and was not completed until the 3d of June, this pier being the last one finished. The excavation at the site of Pier 4 was begun on the 15th September, 1881, but carried on slowly. Pile driving was begun on the 26th November and completed December 27th. Laying of masonry was begun in January, 1882, and the pier was finished on the 12th of May.

The quantities of masonry in the bridge are as follows:

	MASONRY.	CONCRETE.	TOTAL.
East Abutment	70 cubic yards	23 cubic yards	93 cubic yards
Pier I	952 "	779 "	1,731 "
Pier II	2,705 "	847 "	3,552 "
Pier III	2,653 "	860 "	3,513 "
Pier IV	1,090 "	264 "	1,354 "
Totals	7,470 cubic yards	2,773 cubic yards	10,243 cubic yards

More than 7,500 barrels of imported Portland cement and over 3,000 barrels of American cement were used in the masonry of the bridge.

The approach spans are deck trusses of the fish-bellied or inverted bow spring pattern, this form being adopted to keep away from the slope of the embankment. They are entirely of wrought iron, except the pins, which are of steel, and the wall plates, which are of cast iron.

Each of the main channel spans measures 400 feet, divided into sixteen panels of twenty-five feet each. The trusses are fifty feet deep from centre to centre, and twenty-two feet apart. The pedestals, end posts, top chords, and ten centre panels of the bottom chord, and all the pins and expansion rollers, are of steel. All other parts in the main are of wrought iron, except the ornamental work, which is of cast iron. Each long span contains 600,950 pounds of wrought iron, 348,797 pounds of steel and 25,777 pounds of cast iron, the total weight of each span being 975,524 pounds. The steel used was manufactured under the most rigid inspection, and subjected to extraordinary tests before it was placed in position. The extreme height from the bottom of the deepest foundation to the top chord of the bridge is 170 feet. The floor of the structure is formed of oak timbers, nine inches square and fifteen feet long, with spaces of six inches between. On this floor are laid the steel rails of the track.

The east approach to the bridge leaves the old main line at Bismarck Station, and is exactly two miles in length. The west approach is 6,000 feet long from the western end of the permanent bridge, with a descending grade westward of 52.8 feet per mile. One-fourth of this distance consists of a timber trestle, sixty feet at its maximum height, which is built across the space reclaimed from the river by the action of the dike. This trestle spans the place that was the main steam-boat channel of 1880, which is already covered with a fair growth of willows. The bridge was subjected to a severe test on the day it was opened, each span bearing in succession the weight of eight heavy locomotives, or about 520 tons, and the maximum deflection under this enormous burden was not more than three inches.

The Valley of the Upper Missouri.—The fixing of the crossing point of the Northern Pacific Railroad over the Missouri at this place was no accident, but rather the natural result of adequate causes. Long before the remarkable expedition of Lewis and Clark up the Missouri, to its source, over the Rocky Mountains, and down the Clark's Fork and Columbia rivers to the Pacific Ocean, all-sufficing Nature had created and provided for the primitive people of this region a land admirably adapted to their condition and requirements. It was not "a land flowing with milk and honey," but a land abounding in beaver, buffalo and bear, where war and hunting were preferred to money, and life was passed without the white man's cares.

The land that subsisted the countless herds of buffalo, elk, deer, antelope and mountain sheep; that furnished the primitive people of Dakota with the amusement of the chase and the means of supporting a vigorous life, was no "Great American Desert," except in the brains of ignorance, but was really one of the most productive areas of North America. Proof of this productiveness is found not only in the almost limitless prairie growth of sweet and nutritious herbs and grasses, but also in the forests of valuable timber that are found in every place where the streams or bluffs afford protection from the devastating annual prairie fires.

The region for a hundred miles above and below Bismarck and Mandan is blessed with an abundance of this kind of protection. The Big Heart and the Little Heart, on the west of the Missouri, empty their pure and constant waters into the great river very nearly opposite Bismarck—the former just above Fort Abraham Lincoln, and the latter just below. On the east side of the Missouri, Apple and Burned Creeks, streams of great length, purity and durability, flow into the great river just outside of the same town. In addition to these rivers, there are, to the west of the Missouri, the Cannon Ball, the Square Butte, the Knife and the Little Missouri, and to the east the Turtle, the Long Lake, the Horse Head, and the Big and Little Beavers, all affording not only the finest arable and grazing lands, but, from their diffusion of an abundance of pure, fresh water, a protection from prairie fires to an immense tract of heavily timbered country. This timber, from primitive times up to the advent of civilization, gave adequate shelter, during the cold and inclement seasons, to the game that was reared and subsisted on the surrounding prairies, attracting the primitive inhabitants of this region as to a great centre abounding with the means of subsistence. Another peculiarity aided in fixing this point as one of great local interest to the nomadic aborigines. It is the narrowest point on the Missouri River for thousands of miles, and so offered the most feasible and expeditious crossing place for both Indians and the wild herds upon which they lived.

The copious spring and summer rains that contributed to the luxuriant growth of the prairie grasses for the subsistence of the buffalo, also made it possible to raise crops of maize or Indian corn, which were cultivated extensively by the squaws on the rich, sandy alluvion of the Missouri bottoms. That the culture of this grain had been carried on by the aborigines from a very remote period, is shown by the fact that numerous fossilized and many charred corn cobs, in a perfect state of preservation, are still found in the excavated bluffs along the river, and very deep down in the oldest mounds.

Another marked peculiarity of the Missouri Valley in and above the region about Bismarck is its equable and agreeable winters. These result from the warm air which is brought by the westerly winds that prevail in this latitude from off the heated water of the great Japan Current in the Pacific Ocean. The current of air passing eastward over the Northern Pacific Ocean, in its course onward across this part of the continent, finds comparatively little obstruction in the intervening mountain ranges. These were so denuded during the glacial period of the vast quantity of material which was toppled over from their highest peaks and deposited as rich surface soil upon this wheat-growing region, as to render them comparatively very low. The highest point of the Northern Pacific Railroad is less than 6,000 feet, and this is through and over an isolated range, while the Union Pacific Railroad is built for nearly 1,000 miles upon and over ranges of nearly 8,000 feet in height. This low elevation of the mountains on the line of the Northern Pacific allows the heated and saturated atmosphere to pass over with trifling obstruction, thus sensibly ameliorating the climate of Montana and Dakota.

These are a few of the reasons that formerly rendered the valley of the Upper Missouri the "Paradise of Indians." That it was from time immemorial enjoyed as such is easily adduced from what is now known of it. The first recorded observation of it was made by Messrs. Lewis and Clark, who spent the first winter of their celebrated expedition over the Rocky Mountains, in the early part of Jefferson's Administration, at Fort Clarke, situated only a few miles above Bismarck. They give a description of the country and its inhabitants which amply justifies all that is here said of it. The next known of the country was the result of army explorations, and the stories told by the old hunters and trappers of the Hudson's Bay and American Fur Companies. Then came the accounts of the few annual steam-boat voyagers, to whom it was a region very partially explored, and who characterized it as a land of Indians, buffalo, elk and all other kinds of game, with plenty of timber.

In 1863, after the Sioux had perpetrated their unparalleled massacre in Minnesota, General Sibley drove them from that State and followed them to their paradise of subsistence and safety in the neighborhood of the place where Bismarck now stands. Here the red men crossed the Missouri, and the pursuit into the unknown land beyond was abandoned. This expedition, like almost every other against Indians, was a very hard one, and most of the volunteers from Minnesota remembered the country only through their prejudices caused by the hardships of the campaign, which naturally resulted in giving it a bad name. This fact, however, did not deter the Northern Pacific Railroad Company from sending out experienced engineers to find and locate the most feasible route for the railroad, and to select the most practicable place for the construction of a bridge across the broad and swift Missouri River. The first efforts of these engineers were in pursuance of a suggestion made by General I. I. Stevens, upon which they directed their explorations along a line passing a long distance north of the present crossing. When, however, the road, through legislation, had attained the importance which induced its friends to work earnestly for its completion, the engineers, in looking for the most direct and feasible route, encountered the Indian *travois* and buffalo trails leading to this Indian paradise. Following the footsteps of these experienced and successful guides, they followed as far as practicable these trails, and were thus directed to the best crossing place on the river.

The establishment of the crossing here led to the building of Bismarck and the settlement of the surrounding region. This resulted in so thorough an exploration as to furnish all the data necessary to establish the true, natural and inherent value of the country. Everywhere were seen the carcasses of buffalo slain by the Indians, and the ground was found literally cut up by the trails of these animals and the red men who hunted them, leading from all points of the compass on the east side of the river to the present site of Bismarck as a great converging place. All along the river banks, above and below, ancient as well as modern mounds were found, containing Indian skeletons, implements of war and the chase, with specimens of pottery and other evidences of aboriginal ingenuity.

No wonder, therefore, that this region should be the red man's favorite resort, and that they lived here in large numbers. But the capability of the region to subsist men and animals is not only deducible from the footprints of the former inhabitants, but also from the gradual experience of the last five years, during which period the adaptability of the country to support a numerous and prosperous population has been fully demonstrated.

Missouri Division.

MANDAN TO GLENDIVE.—DISTANCE, 216 MILES.

Mandan—(200 miles west of Fargo and 474 miles distant from St. Paul; population, over 2,000)—lies on the western bank of the Missouri, nestled in the lowlands between that great stream and the Heart River, just after the railroad bridge is passed. The city is the county-seat of Morton County. On three sides it is enclosed by low ranges of hills, and the fertile Heart River Valley here broadens into a wide circular plain. Up to 1879, when the extension of the railroad west of the Missouri River was begun, the site of Mandan was occupied by Indians, while buffaloes ranged on the neighboring hills. Even as late as the period named the warlike Sioux had here a series of skirmishes, which culminated in a pitched battle with the Arickarees, or Rees, as they are commonly termed, a branch of the Mandan tribe. Under shelter of the bluffs Mandan was founded two years ago, and its rapid growth has surpassed the most sanguine expectations of the men who determined that a city should be here established. The streets are laid out in squares, the principal thoroughfare being Main Street, which runs parallel with the railroad, but is divided from the track along its entire length by a wide open space that is set apart for a city park. The present number of buildings exceeds 350, and many more are in course of construction. These structures, generally of a substantial character, include large business houses, several churches, good schools, a public hall for lectures, concerts and other entertainments, and a fine railroad depot. This being the terminus of the Dakota Division and the beginning of the Missouri Division of the railroad, there are, at Mandan, extensive machine shops, roundhouse, freight buildings, and every other appliance for the transaction of railroad business, a large number of workmen being employed. The aggregate amount of the city improvements, in 1882, was $240,000. The Mandan *Pioneer*, a daily journal, publishing a weekly edition, is ably conducted and alive to the best interests of the city. The *Times*, another weekly newspaper, is also well supported. The First National Bank, with a large capital, occupies a fine building on Main Street. The Inter-Ocean Hotel, admirably equipped and conveniently situated opposite the passenger station of the railroad, is capable of accommodating a large number of guests. Some enterprising citizens of Mandan, wisely recognizing the fact that no surer method of developing the intrinsic merits of the place could be devised than by offering irreproachable hotel comforts to visitors, built the Inter-Ocean as early as 1881. The structure is of brick, three stories high, occupying a square 100 feet in dimensions, and cost $65,000.

In the vicinity of Mandan an abundance of clay, suitable for manufacturing bricks of the very best quality, is obtained, and this industry is quite prosperous. Excellent stone, also found in the neighborhood, is largely used for the foundation of new buildings. Fuel is supplied in abundance—both wood and coal—by the timber which skirts the rivers and by the mines which are worked to great advantage on the line of the railroad westward. The coal is delivered by the car load at the low rate of $3.25 per ton. The outlying lands are very fertile, and large crops of wheat, corn, potatoes and other vegetables are produced. Much attention is given in Morton County to stock and sheep raising, to which the country and climate are well adapted.

With so many solid advantages in her favor, it is scarcely a matter of surprise that the population of Mandan nearly doubled during the year 1882, and that her wealth increased four-fold in a twelvemonth. Her merchants not only do a large business with the farmers who are flocking into the fertile regions westward, but also send supplies by steam-boat to the posts and settlements of the upper Missouri, the fine rock landing on the river at Mandan affording a peculiarly favorable point of shipment. The taxable valuation of the city in 1881 was about $160,000. This had increased in 1882 to $600,000, according to the Assessor's returns in May of that year, and in the spring of 1883 these figures had advanced to $1,000,000.

Near Mandan are points of interest dating from prehistoric times. A short distance south of the city are mounds which have been formed by successive layers of camp refuse, heaped together and burned by recurring prairie fires. In these stratifications are found stone weapons, arrowheads, household implements, pottery, trinkets and bones of men and animals. The Indians deny all knowledge of these mounds, the presence of which offers a fine field for archæological and ethnological research. The Mandan *Pioneer* of April 27th, 1883, printed the following in relation to this subject:

"Two miles from Mandan, on the bluffs near the junction of the Heart and Missouri Rivers, is an old cemetery of fully 100 acres in extent filled with bones of a giant race. This vast city of the dead lies just east of the Fort Lincoln road. We have just spent a half day in exploring this charnel house of a dead nation. The ground has the appearance of having been filled with trenches piled full of dead bodies, both man and beast, and covered with several feet of earth. In many places mounds from eight to ten feet high and some of them 100 feet or more in length have been thrown up and are filled with bones, broken pottery, and vases of various bright colored flints and agates. The pottery is of a dark material, beautifully decorated, delicate in finish and as light as wood, showing the work of a people skilled in the arts and possessed of a high state of civilization. Here is a grand field for the student, who will be richly repaid for his labors by excavating and tunneling in these catacombs of the dead. This has evidently been a grand battle field where thousands of men and horses have fallen. Nothing like a systematic or intelligent exploration has been made, as only little holes two or three feet in depth have been dug in some of the mounds, but many parts of the anatomy of man and beast, and beautiful specimens of broken pottery and other curiosities, have been found in these feeble efforts at excavation. Who are they and from whence did they come, dying, and leaving only these crumbling bones and broken fragments of their works of art to mark the resting place of a dead nation? Five miles above Mandan, on the opposite side of the Missouri, is another vast cemetery as yet unexplored.

"How long have these bones and remains laid in this cemetery? is a question which readily suggests itself. The fact that there are no existing tribes on the plains having any knowledge of pottery would indicate that the mounds had existed for a very long time. And yet there are found near the surface, and again down to a depth of nine, ten or fifteen feet, well preserved bones, which look as if they had not been buried more than five or ten years. Then again, the fact must be borne in mind that there are no tribes existing that will own to any knowledge of these mounds. The Indians simply say they are spirit mounds, concerning which they know nothing. It seems strange that they should have been forgotten even within a period of 100 or 200 years, since the Indians have very tenacious memories for traditional matters. The sexton of this cemetery appeared to have a very peculiar way of doing his work. It seems that human bodies were buried, then an accumulation of grass and brush was thrown over them and set on fire. This is proved by the fact that above the bodies will be found from two to three inches of ashes. Then it looks as if the living folks had remained in the vicinity long enough to cover the dead remains with broken pottery and bones of animals. The whole would then be covered with layers of rubbish, such as would be cleared away from the tents of the people as a sanitary precaution. Broken pottery, and fragments of bones and ashes in layers, go to make the funereal mounds complete.

"In the ashes are found charred corn-cobs, burned bones and charred meat. All the large bones that are found are broken, with the exception of the human bones. Judging from appearances, this was not only a great cemetery, but a great banqueting place also."

A Legend of the Heart.—In the newspaper before quoted, under date of May 4th, 1883, the editor has something further to say with reference to the remarkable burial place. In rambling over the bluffs he encountered an aged Mandan Indian, named Red Bird, wrinkled and bent with years, but unusually intelligent for one of his race, who narrated the following legend, which is worth reproducing, although it may be nothing more than a lively bit of romance:

"Many moons before the red man lived in the land of the Dakotas, a powerful tribe or race of people lived on the banks of the Heart River. They were skilled in the arts of war and the chase—a nation of giant men and beautiful women, living in tepees built like the pale faces make their houses. This people, like the leaves of the forest, could not be numbered, possessing large herds of horses, cattle and other animals. All other tribes were subdued by them and became their servants. They went mounted on powerful and fleet horses, armed with huge spears and bows and arrows, and had large quantities of gold, silver, beautiful flints and precious stones, and made handsome vessels of burnt clay, decorated with flowers and animals. This people, like the white man, worshiped the Great Spirit in grand tepees erected in their cities, filled with gold and silver and beautiful vessels made from bright colored flints and agates. They were ruled by a chief, or king, who was distinguished for his valor in war—a mighty giant in size—whose will was law, and who encouraged the works of art and science and established great schools of learning and athletic games among his people, and, from a grand amphitheatre, filled with beautiful women and brave men, distributed prizes among those excelling in the arena. They were a grand, proud, happy people, and lived many years on the banks of the Heart.

"But a powerful nation of dark-skinned men came from the North and made war on them. Many battles were fought and thousands slain on both sides. Finally a great battle was fought at the mouth of the Heart, lasting many days, in which these dark-skinned men surrounded the people of the Heart, killing every man, woman and child, piling up their dead bodies in trenches, pulling down their grand tepees on them, setting all on fire. After feasting many days, celebrating their victory, a pestilence broke out, a spotted disease, which destroyed the last man among them, leaving not one man, not one woman, nor child, nor four-footed beast of any description whatever, over this vast extent of territory, to mark the victor or vanquished. Only bones, ashes, broken pottery and a few implements of war in these little mounds were left to tell the tale of the once happy and powerful people of the Heart.

"Nothing more; and Red Bird, too, will soon pass to the happy hunting grounds, almost the last of his once powerful race."

Marmot.—After leaving Mandan the railroad passes through the fertile Valley of the Heart River, which tortuous stream it crosses at frequent intervals, before reaching Marmot, the next station, nine miles westward. Marmot is situated on a high plateau, near the confluence of the Heart and the Sweetbriar rivers. The station derives its name from the fact that a prairie dog village existed here before the railroad appeared. As the train advances westward these curious little animals are more abundant, their antics affording a great deal of amusement to passengers. Col. Richard I. Dodge, in his book "The Plains of the Great West," writes that "this well known animal is badly named, having no more of the dog about him than an ordinary grey squirrel. He is a species of marmot and burrows in the ground as do wolves, foxes, raccoons, skunks and all the smaller animals on the treeless plains. He lives on grass and roots, and is exceedingly prolific, each female bringing forth several sets of young each year. He is not excellent eating, but the young are as good as the common squirrel, and, when other flesh meat is not to be had, they make no unwelcome addition to the bill of fare. I regard the prairie dog as a machine designed by nature to convert grass into flesh, and thus furnish proper food to the carnivora of the plains, which would undoubtedly soon starve but for the presence in such numbers of this little animal. He is found in almost every section of the open prairie, though he prefers dry and arid to moist and rich localities. He requires no moisture and no variety of food. The scanty grass of the barest prairie appears to furnish all that is requisite for his comfortable existence. Though not in a strict sense gregarious, prairie dogs yet are fond of each other's company, and dig their holes in close vicinity. Such a collection is called a town, and they sometimes extend over immense areas. The numbers of inhabitants are incalculable. Cougars, panthers, wild cats, wolves, foxes, skunks and rattlesnakes all prey upon them without causing any perceptible diminution of their immense numbers."

Sweetbriar—(16 miles west of Mandan.)—This is a new settlement, situated on the river of the same name. There is here a section house, and the locomotives are supplied with water. A few miles south of this point is an extensive sheep ranch, sheep husbandry being profitably carried on at many other points also, and promising to develop into a great business in Dakota.

Topographical.—For 100 miles westward the physical appearance of the country is that of a roughly rolling prairie, the fine agricultural possibilities of which have already been successfully tested. The railroad crosses at frequent intervals many water courses, the more important of which, after leaving the Heart, are the Curlew and Knife rivers and Beaver Creek. These streams are no puny rivulets, but dignified rivers of considerable volume, which, with their tributaries, meander in devious ways throughout the length and breadth of the land grant of the railroad, forty miles on either side of the track. Along these water courses there is usually a fair supply of soft wood timber, and the land is everywhere covered with a rich growth of buffalo and other nutritious grasses. The horizon is bounded on all sides by the undulating outline of the surface, varied occasionally by some dominating elevation which serves as a landmark. These sharp, conical elevations, denominated buttes,* are very peculiar. They rise from the rolling plains, and, being usually without vegetation, show the sedimentary strata of the soil, which is often of many colors. All this region is at present thinly inhabited; but, as it is endowed with good water, an abundance of lignite coal, a rich soil and a climate even somewhat milder than the country eastward, its advantages for settlement have been already recognized.

Sedalia—(24 miles west of Mandan.)—This is a side track, on the summit of the Sweetbriar, established in the midst of a good farming and coal mining district. *New Salem*, the first station west of Sedalia, is a new place which is likely to grow.

Blue Grass—(32 miles west of Mandan), with a section house and side track. Settlers are fast taking up the land between this place and Sims, the next station.

Sims—(35 miles west of Mandan; population, 150.)—This place is in the midst of a good coal region, and the people are mainly engaged in mining. It is named in honor of Mr. George V. Sims, chief clerk in the executive office of the Northern Pacific Railroad in New York. The coal mine has a shaft fifty-six feet in length, with sloping cuts into the bluff 1,000 feet long. There are four veins from three and a half to four feet in thickness, and one vein of seven feet. The output, over 100 tons per day, is a superior quality of lignite, which finds a ready market on the line of the railroad. The coal is abundant, and measures have been taken to increase the capacity of the mine by sinking a new shaft and doubling the working

---

* In pronouncing this word the *u* is sounded as in tube.

force. Building operations are quite active in the town, and lumber mills, brick yards and other manufacturing interests are to be established. In the vicinity of Sims an excellent clay is found in abundance, which is not only well adapted to brick making, but also to the manufacture of terra cotta work and all kinds of fine pottery. Terra cotta works and brick making machines are already in operation, the new school-house being constructed of pressed brick, with terra cotta trimmings, which were manufactured in the town.

Curlew—(46 miles west of Mandan)—pleasantly situated in the Curlew Valley, has a section house and side track. The valley, ranging from two to four miles in width, has a light colored, sandy soil, containing sufficient alkali to produce large crops of wheat and oats.

Kurtz—(53 miles west of Mandan)—has a section house and side track. As illustrating the remarkable fertility of the soil, it may be said, upon the authority of the railroad superintendent, that this place remained the end of the track during the winter of 1880-1881, and numbers of horses were fed here. Quantities of oats were naturally scattered over the ground, the result being a spontaneous growth upon the sod, from which was harvested in the proportion of at least sixty bushels to the acre.

Glenullen is a new station west of Kurtz. It was settled in the spring of 1883 by a colony of farmers and mechanics from Ohio and Wisconsin, numbering 100 or more persons. The place has inviting features, and is likely soon to be a thriving town.

Eagle's Nest—(63 miles west of Mandan)—is a fuel and water station, with a section house for the track men. The water is conducted to the tank, without the necessity of pumping, through a pipe from an elevated spring three-quarters of a mile distant.

Knife River—(72 miles west of Mandan.)—This station is situated on the Big Knife River, a stream larger than the Heart at Mandan, which pursues its way north through a beautiful valley, until it finally empties into the Missouri. There are here a section house, water tank and side track. *Antelope*, seven miles beyond, is also provided with similar facilities for carrying on the work of the railroad.

Richardton—(86 miles from Mandan)—is a new place, founded only in the autumn of 1882, and named in honor of Mr. C. B. Richards, of the firm of C. B. Richards & Co., of New York, passenger agents of the Hamburg

Steamship Line. The town is situated in Stark County, near Young Man's Butte, a prominent elevation not far from the railroad, and the promoters of the place have already succeeded in giving it importance. There are a number of stores, an hotel, a lumber yard, a brick yard, an elevator, a creamery, and building operations are active to supply the needs of a rapidly growing population. The surrounding country rolls in regular undulations through miles and miles of fertile soil, offering superior advantages for farming. The soil is a dark, rich and somewhat sandy loam of great depth, underlaid with a clay subsoil, and is well adapted to the cultivation of wheat, rye, oats and barley. To the north of Richardton the country is somewhat broken, interspersed with well watered valleys that afford abundance of wild hay. The small streams are generally fringed with a growth of cottonwood trees, thus making the region admirably suited to successful stock and sheep raising. Inexhaustible beds of coal, which may be inexpensively mined, underlie the whole region.

Taylor—(91 miles west of Mandan.)—This town is surrounded by a wide expanse of fertile country. The soil is of vegetable mould, eighteen inches to three feet deep, with a fine subsoil similar to that of the James River Valley. Four miles south of Taylor flows the Heart River, while to the north is the Big Knife. Both these streams have broad, grassy valleys skirted with groves of oak, cottonwood and ash. Here, too, are found excellent cattle and sheep ranch sites. Many springs of good water issue from the outcropping beds of coal in the bluffs bordering the valleys, and wells give a good supply at a depth of sixteen to thirty-five feet. Besides the fuel which is furnished by the oak and cottonwood trees, the whole country is underlaid with a bed of good coal five feet in thickness, which can be mined by digging from three to fifteen feet deep. From this bed the settlers obtain their own fuel at leisure times, highly appreciating so great an advantage. Taylor has several stores and an hotel.

Gladstone—(98 miles west of Mandan; population, 300.)—This town was laid out in the spring of 1882 by a colony from Ripon, Wisconsin, near the fertile valley of the Green River. The situation of the town is pleasant and the surrounding country for many miles is settled by the colonists. During the first year of the colony's existence about 150 families took up the lands in the neighborhood, and the crops raised upon the upturned sod were bountiful. Near Gladstone are great fields of coal of a good variety for heating and cooking purposes. This coal is apparently of a recent formation, and emits no smoke or disagreeable odor, but burns like wood and equally as fast. Gladstone has an hotel and a number of stores and shops.

Dickinson—(110 miles west of Mandan; population, 400)—is a bright new town in the valley of the Heart River, at the terminus of the first freight division of the Missouri Division. It lies in the midst of an agricultural and grazing country, and promises to become a great shipping point for cattle and grain. The ground on the outskirts of the town gradually slopes to the south, giving a fine opportunity for drainage. The buildings are of a permanent order, superior in appearance and construction to those usually found in new towns. There are a good hotel, a large general store, a fine bank building, church organizations and schools, with commodious railroad shops, round-house, passenger depot and freight warehouse. Dickinson will doubtless be the county-seat of Stark County. The tributary country is well watered, and the rain-fall in spring and summer is sufficient to ensure good crops. Many thousands of acres are already under cultivation, and there are excellent stock ranges within thirty miles of the town. The coal beds in the immediate vicinity produce a good quality of lignite, and a fine grade of clay for brick making and sandstone for building purposes is found in the neighboring bluffs.

It was at Dickinson that a gentleman from the East, in conversation with an old hunter and trapper, asked what kind of Indians the Sioux were, and if they could be trusted. The old hunter answered that he hardly knew, as he had been roving around these parts for nearly ten years and remembered only one Indian whom he really cared to trust. The Eastern gentleman thought it rather strange that there was but one among so many, and asked where this good Indian had been seen, to which the hunter replied that he had seen him hanging to the limb of a tree. A smile covered the old man's face as he said, "Good day, neighbor; I'm going."

South Heart—(121 miles west of Mandan.)—This station has a water tank and a section house. The soil is productive and farm houses are fast dotting the landscape.

Belfield—(130 miles west of Mandan)—is situated in a region which is sometimes termed the "Summer Valley." The Heart River, here a pretty stream, is bordered on each shore with handsome trees. Hundreds of miles north and south of the new town stretches a very fine agricultural country, and its proximity to the well sheltered valleys of the Bad Lands will make it a head-quarters for cattle raisers. Belfield contains a church, several general stores, a fine depot and lumber and lime yards. A banking company has been formed, and a flouring mill, a grain elevator and an hotel are to be established. Excellent clay for brick making purposes is obtained in the immediate vicinity, and a brick yard is in operation. Water is found in abundance by digging wells at no great depth. The lignite or soft coal which underlies the whole section will furnish ample fuel to the settlers.

The next stations—*Fryburg*, *Sully Springs* and *Scoria*—are in the midst of scenes which are so unique as at once to fix attention.

The Bad Lands.—At Fryburg the train suddenly leaves the beautiful rolling prairies and enters a long cut on a down grade, presently emerging upon a region, the startling appearance of which will keep the vision alert until the Little Missouri River is reached, fourteen miles beyond. Here are the Bad Lands, sometimes called Pyramid Park, which show that the mighty forces of water and fire, fiercely battling, have wrought a scene of strange confusion. Buttes, from fifty to 150 feet in height, with rounded summits and steep sides, variegated by broad, horizontal bands of color, stand closely crowded together. The black and brown stripes are due to veins of impure lignites, from the burning of which are derived the shades of red, while the raw clay varies from a dazzling white to a dark gray. The mounds are in every conceivable form, and are composed of different varieties of argillaceous limestone, friable sandstone and lignite lying in successive strata. The coloring is very rich. Some of the buttes have bases of yellow, intermediate girdles of pure white, and tops of deepest red, while others are blue, brown and gray. There are also many of these elevations which, in the hazy distance, seem like ocean billows stiffened and at rest.

Between these curiously shaped and vari-colored mounds there are sharp ravines and gulches which are often the beds of shallow streams. Here and there are broader spaces, covered with rich grass and flecked with a growth of ground juniper of delicious fragrance. No trees worthy of the name are seen, but a fringe of gnarled and misshapen pines occasionally presents itself along the water channels. In ages long ago, however, dense forests existed in these Bad Lands. There is evidence of this primeval growth in the abundant petrifactions of tree stumps, four to eight feet in diameter, which are in portions translucent as rock crystals, and susceptible of as high a polish. Fine specimens of fossil leaves, of the Pliocene age, changed by the heat of the burning lignite into a brilliant scarlet, but retaining their reticulations perfect, are also found. The coal,

Buttes in Pyramid Park.

still burning, gives a Plutonic aspect to the whole region, one fiery mass not far from the railroad being easily mistaken at night for an active volcano, the cliffs having close resemblance to volcanic scoria. Among the many other fossil remains are oysters, clams and crustaceans. The seeker for geological curiosities has here a fine field in which to work.

The term Bad Lands, as applied to this region, is a gross misnomer. It conveys the idea that the tract is worthless for agricultural or stock raising purposes. Nothing could be wider of the truth. The fact is, the soil possesses fertilizing properties in excess, and the luxuriant grasses which here flourish attract herbivorous game animals in large numbers. The designation "Bad Lands" is derived from the times of the old French *voyageurs*, who, in their trapping and hunting expeditions in the service of the great fur companies, described the region as "*mauvaises terres pour traverser*," meaning that it was a difficult region to travel through with ponies and pack animals. This French descriptive term was carelessly translated and shortened into "bad lands," and thus has resulted a wholly false impression of the agricultural value of the country.

This entire region, geologists tell us, was once the bed of a great lake, on the bottom of which were deposited for ages the rich clays and loams which the rains carried down into its waters. This deposit of soil was arrested from time to time sufficiently long to allow the growth of luxuriant vegetation, which subsequently decayed and was consolidated by the pressure of succeeding deposits, transforming itself into those vast beds of lignite coal, which abundantly meet the need of the country for fuel. The various strata thus deposited are all of recent origin, and, being without cementing ingredients, remain soft and easily washed by the rains. When at last this vast lake found an outlet in the Missouri, the wear and wash of these strata, under the action of rain and frost, were very great. Hence the water courses, especially the minor ones, where the wash has not had time enough to make broad valleys, have precipitous banks and high, enclosing bluffs, with curiously furrowed and corrugated sides, usually bare of vegetation, and showing only the naked edges of the rich soils of which they are composed. The tops of these bluffs and buttes are on the general level of the whole country, and are equally as fertile. This is shown by the hotel garden at the Little Missouri, where, in the very heart of the "Bad Lands," and on the summit of the highest bluff, a level spot was chosen and planted, which annually yields heavy crops of vegetables, the potatoes alone producing as many as 300 bushels to the acre. But these Bad Lands, misnamed as they are, form a very small part of the country—they are conspicuous from the fact that the chaos of buttes is so curious and fantastic in form and beautiful in varied color. From the railroad, which naturally follows the valleys between these strangely formed, isolated mounds and hills, the view of the broad, open country which lies on a level with their tops is shut off.

The writer made a short excursion into these "Bad Lands" in the autumn of 1882, and noted his observations in an Eastern journal as follows:

"My visit to the Bad Lands—which, by the by, are beginning now to be known as Pyramid Park—proved to me how erroneous had been my own impressions with respect to them. I found excellent grazing in all the tortuous valleys and frequent glens; while the tops of the giant buttes—level as a floor in many cases, and containing hundreds of acres in a single plot—offer as fine agricultural lands as can be found.

"A party of six arrived late at night at the hostelry at Little Missouri Station, a rough, but not uncomfortable refuge for tired and hungry wayfarers. After a good night's rest we started next morning on a tour of exploration, guided by Moore, the inn keeper, a jolly, fat and rosy-cheeked young man, brimming over with animal spirits. Two of the party preferred riding on a buck-board wagon; the others mounted hardy 'cayuse' ponies, and among the latter was a subject who weighed 250 pounds, his avoirdupois fully testing the wiry endurance of his steed, which showed no sign of flagging vigor after a long day's journey. Twice we forded the shallow stream, yellow as the Tiber. Rough riding here. If I were to tell of the slopes down which we slid, and up which we struggled, buck-board and all, I am sure I would jeopard my reputation for truth. The ride was quite exhilarating and altogether novel, nevertheless. A particular zest and flavor was given to the scenery by the remarkable grouping of fantastically shaped buttes, each girdled with a broad band of crimson—a stratum of pure pottery, burned in Nature's oven by the combustion of the coal veins underlying the clay. These potsherds, jagged and shapeless, are used by the railroad instead of gravel for its road-bed, and answer the purpose admirably. The road here, therefore, may well be likened to a scarlet runner. After some hours of rough riding we brought up at a sheep ranch, belonging to the Eaton Brothers, where we were surprised by the many appliances for comfort en garçon. A tame antelope fawn, playful as a kitten, and a medley of buffalo heads and elk and mountain sheep horns, as well as other trophies of the chase, diverted us, and stories were told of the large and small game which the neighborhood supplies to those who know how to shoot it, which would make even the least enthusiastic sportsman long to try his luck. A lavish game dinner, including tender buffalo steak, washed down with rich milk and good water, and a dessert of canned fruits, was just the thing to satisfy appetites made unusually keen by the brisk ride in the dry, pure air. These Eaton boys, whose hospitality we had so agreeably tested, are from the East, and they have money enough invested in sheep and cattle to carry on a very respectable wholesale business in any large city of the Union. Having enjoyed their hospitality as long as our time would admit, we left their 'shack,' which is the common name

Pyramid Park Scenery.

for a substantial log house, reinforced by one of these happy ranchmen—a young chap who sat his horse as though he were a centaur, and looked a picturesque and noble figure, with his clean shaven cheeks, heavy drooping mustache, sombrero, blue shirt and neckerchief with flaming ends; in fine, a perfect specimen of the noble manhood finish which this breezy, bounding Western life often gives in a few years to the Eastern born and bred young man. After visiting a coal vein which has been smouldering constantly ever since the country was known to the whites, and from time immemorial, according to Indian tradition, the fire of which is visible at night from the train, we inspected the 'Maiden of the Park,' the 'Watchdog,' and others of the buttes which bear more or less resemblance to the things after which their sponsors named them. We also chipped off specimens of petrified wood, full of sparkling, silicious crystals, from the mammoth tree trunks turned to stone, which crop out from the sides of the conglomerate mounds, showing that, in ages long remote, a stately forest grew on these grassy plains."

Professor N. H. Winchell, of Minnesota, who accompanied General Custer as geologist on his Black Hills expedition in the summer of 1874, thus describes the general formation of this region:

"Although I call these bad lands (for so they are generally known among the men who have before crossed here), they are not so *bad* as I had been led to expect from descriptions that I have read. There is no great difficulty in passing through them with a train. There are a great many bare clay and sand buttes, and deep perpendicular cañons, cut by streams in rainy seasons; but there are also a great many level and grassy, sometimes beautiful valleys, with occasionally a few trees and shrubs. There is but little water in here, the most that we have found being due to recent rains. The tops of a great many of the buttes are red, and often they are overstrewn with what appears like volcanic scoria. This, I am satisfied, arises from the burning of the lignite, which occurs in nearly all these lands, there being one large bed of it, and sometimes two distinct beds, in the same slope. The lignite is ignited by fires that sometimes prevail over the plains, set by Indians, and when fanned by the strong winds that sweep across them, produces a very intense heat, fusing the over and underlying beds, and mixing their materials in a confused slag, which, although generally of a reddish color, is sometimes of various colors. The clay makes a very hard vitreous or pottery-like slag, that is sometimes green or brown. Iron stains the whole with some shade of red."

Little Missouri—(150 miles west of Mandan; population, 100.)—This station, on the Little Missouri River, snugly situated in a valley plateau, is known as Comba Post-office. It contains general supply stores, which are well supported by the sheep and cattle ranchmen who occupy ranges within a few miles of the station. A company of soldiers occupied a cantonment here during several years for the protection of settlers against the Indians, but the Government abandoned the post in the spring of 1883, and sold the buildings, which have been converted into an hotel called the Pyramid Park Hotel. The house is under the management of Mr. Frank Moore, a pleasant and jolly host, who knows the entire region thoroughly, and is an experienced hunter. Several extensive cattle and sheep ranches established on the line of the river are doing well, this being an excellent grazing point. The Marquis De Mores, a retired officer of the French army, has scripted some 200 acres of land on the east and west side of the river at Little Missouri, which is to be the scene of extensive packing and slaughter house operations. He intends to carry on sheep husbandry on an extensive scale. His plan is to engage experienced herders to the number of twenty-four, supply them with as many sheep as they may desire, and provide all necessary buildings and funds to carry on operations for a period of seven years. At the end of this time a division of the increase of the flocks is to be made, from which alone the Marquis is to derive his profits. Many of the sheep have already been purchased and are on the various ranches. The Marquis has also organized the Northern Pacific Refrigerator Car Company, and associated with him a number of Eastern stockholders, the intention being to slaughter and pack at the Little Missouri station, cattle, sheep, hogs and game. Lignite coal is mined to a considerable extent here, and used by settlers and the railroad company. Little Missouri is a capital place at which to fit out hunting excursions. There are elk, deer, antelope, big-horn or mountain sheep and buffalo to be secured in all the region, and experienced men may be engaged as guides.

Frontier Manners.—An amusing incident illustrating frontier manners occurred at Little Missouri in the early days of the railroad, when Mr. Moore kept only an eating house. An Eastern "tenderfoot" stepped in from the train one evening to partake of a meal. Seating himself at the table, his supper was brought in. The waiter in attendance, a lad of perhaps fifteen years (a young cowboy in disguise,) considered it would be the polite thing to give the gentleman his choice of tea or coffee, thinking that the guest would notice that he held the coffee pot in his hand and would, of course, take coffee. He therefore called out: "Tea or coffee, sir?" The gentleman, without looking up, answered: "I will take tea, if you please." The boy flashed into a rage, and sternly said, with an oath: "*You will take coffee*, or I will scald you!" The astonished guest looked up quickly and said: "*I will take coffee.*" The young man had tea in the kitchen, but thought it was too far to go for it. Another story is that of the fastidious "tenderfoot" who had spread his blankets and rested as well as possible in a room with a score of others in one of the rough lodging places of a budding frontier town. He was at a loss for a place where he could perform his ablutions, but at last found the common lavatory. A coarse and grimy towel hung upon the door on a roller. Loath to use so uninviting a rag, the guest besought the landlord to give him something cleaner. At this the latter was mildly astonished, remarking: "Look here, stranger, thirty men have used that towel this morning, and nobody has complained of it but you," adding, with a deadly gleam in his eye: "We don't allow grumbling in this hotel."

Soon after leaving the Little Missouri River the country westward becomes less rough, although the railroad passes through many cuts and ravines. Gradually, however, the feature of the landscape is that of broad rolling prairie marked here and there by isolated buttes. The last two stations on the railroad in Dakota are *Andrews* and *Sentinel Butte*, distant respectively 158 and 166 miles from Mandan. These places are both unimportant.

Sentinel Butte is a prominent object on the left hand, not far from the track. The top of this eminence is visible on clear days at a distance of thirty miles, but looks only to be about three miles off, so deceptive is the luminous atmosphere. This region abounds in moss agates, specimens of which are found, near the foot of the buttes, of great size and beauty. A well known army officer, who was at one time stationed here, secured a sufficient number of these agates, so large that they were converted into dessert knife handles, and served as a unique and handsome present to a lady on her wedding day.

Sentinel Butte, in spite of its precipitous faces, as seen from the railroad, is easy of access on the side remotest from the track. On its summit there is half an acre of level ground. Buffalo were very partial to this elevation, and sometimes resorted to it in so large numbers, that many were crowded over the brink. The bones of these animals lie in heaps at the foot of the precipice, whitened by the weather.

A Shot at the Buffaloes.—At various points on the line of the railroad there is still a chance to get sight of buffaloes which have strayed from the main herd away to the north. This is a piece of good fortune of which passengers are especially proud, and formerly it was no infrequent experience. But the opportunity of seeing buffaloes and shooting at them from the train grows rarer year by year. When, however, the animals do show themselves near the track, be sure that their presence will be made known by a fusilade from the car windows. The buffaloes are usually seen scampering off in single file toward the shelter of the distant bluffs, behind which they soon disappear. The episode is always exciting, but neither man nor beast is hurt. In a very short time no trace of the millions of buffaloes which recently roamed over these Western plains will exist. They are slaughtered by the hundred thousand every year—in winter for the robes, in summer for the hides. Numbers of men follow buffalo killing as a regular occupation, and think their calling just as legitimate as that of building houses and tilling the soil. The great southern herd of buffalo has already been exterminated, and the northern herd alone remains. This latter herd now ranges over an area about 350 miles in diameter in north-eastern Montana and the Northwest Territory. In 1882 at least 200,000 were slaughtered, and this year will probably add as many more victims to the destroyer. At some stations on the railroad, in the Yellowstone Valley, many thousand buffalo hides are usually piled up like cord wood for shipment to the East, and a number of hunters, of fierce and frontier mien, will be seen guarding them. But, after buffalo meat becomes very scarce, the Indian will, perhaps, find it needful to raise his own beef supply. He will be likely in that case to settle down to agricultural and other civilized pursuits, and so develop into an enterprising and useful person. Here comes in the law of compensation. The buffalo is to be lamented, but the Indian will be redeemed.

A Primitive Boundary Mark.—One mile west of Sentinel Butte the boundary between Dakota and Montana is crossed. The line is marked by a tall pole, upon which is nailed a fine pair of antlers.

The railroad for the next thirty miles passes over a fine prairie plateau which is watered by many small running streams. It then traverses six miles of broken country which form the divide between the Little Missouri and Yellowstone rivers, after which it descends into the valley of Glendive Creek and reaches the Yellowstone River at the town of Glendive, twelve miles beyond.

Montana.

Montana embraces nearly as large an area as Dakota. It averages 275 miles from north to south, and 550 miles from east to west, stretching through 12° of longitude, from 104° to 116° west of Greenwich, and lies for the most part between the forty-fifth and forty-ninth parallels of north latitude. Its southern boundary is in about the latitude of St. Paul, Minn., and its northern line joins the British Possessions. The mean height of Montana above the ocean level is estimated at 3,900 feet, the greatest elevation among the mountain peaks being 11,000 feet, and the lowest, on the Missouri River, being about 2,000 feet. Of the 93,000,000 acres contained within the limits of the Territory, two-fifths are mountainous and three-fifths valleys or rolling plains. The water-shed between the Atlantic and the Pacific Oceans, the main chain of the Rocky Mountains, traverses the western portion of Montana in a course a little west of north, leaving about one-fourth of the entire Territory on the western slope and three-fourths on the eastern. In the central part of the Territory are the Bull, Belt, the Little Rocky and other smaller mountain ranges, which, with many lateral spurs and detached groups, give that great diversity of rocky ridges, broad plateaus and pleasant valleys, which render the country extremely picturesque.

Montana is well supplied with rivers. Her great water courses are Clark's Fork of the Columbia and the Missouri rivers, the latter with many important tributaries. The Clark's Fork drains 40,000 miles of the Territory, and flows into the Columbia River, while the Missouri and its tributaries, the Milk, the Yellowstone, the Teton, the Marias, the Judith, the Musselshell, the Jefferson, the Madison and the Gallatin carry off the waters of double that area. These rivers are navigated by steam-boats a distance of 1,500 miles within the limits of the Territory. Montana is also supplied with a great number of beautiful lakes, the largest of which are Flathead, in Missoula County, ten by thirty miles in size, and Red Rock, in Madison County, twenty-five miles in length and 6,500 feet above the sea level. The great cataract of the Missouri River, thirty miles above Fort Benton, with a vertical fall of about eighty feet, is renowned for its grandeur.

The agricultural lands of Montana lie mainly in the valleys of the large rivers and their affluents. These valleys, usually old lake basins, which have received the wash from the surrounding mountains, have an alluvial soil which has proved to be very fertile. The land has generally a gentle and regular slope from the higher ground which separates the valleys from the foot-hills, and this is a fact of great importance in its bearing upon irrigation. So uniform is the slope that, in almost every instance, when water is conducted by means of a ditch from any stream, it may be made to flow over every foot of land in the valley below. The uplands (or bench lands, as they are commonly termed) are simply continuations of the valleys at a higher elevation. They frequently look like artificial terraces of enormous size, rising one above the other, and where the quantity of water in the stream above admits the irrigation of the bench lands, they are also found to be very productive. Beyond these terraces are the foot-hills, with rounded tops and grassy slopes, and behind these loom up the mountains, crowned with a scanty growth of pine and fir, although the slopes and valleys are always destitute of these varieties of timber. There are no deciduous trees either, excepting groves of cottonwood and willows along the water courses, and occasional copses of quaking-asp in wet places on the sides of the mountains. Only in the extreme northwestern part of the Territory is a very large body of magnificent timber, covering mountains and plains alike.

Eastern Montana, stretching from the base of the Rocky Mountains to the boundary of Dakota, and embracing an area of 90,000 square miles, is divided into three belts of nearly equal size by the Missouri and Yellowstone rivers. On the west and south are mountains, timbered with pine and fir, and from them issue many streams, which abundantly water the country. The ground is covered with a rich growth of bunch grass, which makes the region an excellent stock range. But the large area of grassy, rolling table lands in the northeastern part of the Territory is pre-eminently the place for cattle raising and sheep husbandry, Meagher County especially, in which lie the Musselshell, the Judith and the Smith rivers, being famed as the great grazing county of Montana.

The resources of the entire Territory are varied and very valuable. Millions of acres of good agricultural land are awaiting development; but, owing to the light rain-fall, irrigation is generally necessary. The improved land reported in 1882 was 516,100 acres, valued for purposes of taxation at

$4,476,118; town lots and improvements, $4,163,618; horses, 67,802, valued at $3,197,000; sheep, 362,776, valued at $1,018,000; cattle, 287,210, valued at $4,699,812; swine, 7,101, valued at $45,249. The total assessed valuation of the Territory was $33,212,319, an increase upon the previous year of $9,170,512, or about twenty-eight per cent. The estimated increase for 1883 is fifty per cent., owing mainly to the extension of the Northern Pacific Railroad and the heavy immigration.

Mining has always been, and probably will continue to be, the leading industry. During the last twenty years the total output of the precious metals in Montana amounted in value to $170,639,843, and the product of 1882 was $8,000,000. A great increase in the yield of the mines must, however, necessarily result from the transportation facilities with which the Territory has only now been favored.

In addition to its bullion shipment, Montana sent out in 1882, 50,000 cattle, 40,000 sheep, 30,000 hides, 40,000 buffalo robes, 100,000 undressed buffalo skins, many thousand bales of elk and deer skins, and 2,500,000 pounds of wool.

Montana was organized as a Territory in 1864, but until 1880 there was no railroad within her boundaries. Her progress in every direction was retarded by this utter lack of methods of transportation, and only now is she beginning to bring forth her latent resources. Under the impetus which has been given by the construction of the Northern Pacific Railroad throughout the entire width of the Territory, and the extension of the Utah and Northern Branch of the Union Pacific from the south, population is fast increasing, and the number of inhabitants already verges upon 100,000, exclusive of Indians upon five reservations, who number between 21,000 and 22,000.

Historical.—The history of Montana has not been destitute of stirring incident. Before 1861 there were no settlements, and the only whites who had visited the region were trappers, missionaries and the members of various military exploring parties. Public attention was first directed to the Territory at about the period named by the discovery of gold in paying quantities in Deer Lodge County. The report brought an irruption of miners from all the Western States, among whom were some of the wildest and most reckless characters, whose names and misdeeds figure in the early annals of the Territory. In 1862 the rich placers at Bannack were discovered. In the following year a party, returning from an unsuccessful attempt to reach the Big Horn Mountains by way of the Gallatin River, whence they were driven back by the Crow Indians, camped for dinner on Alder Creek, near the site of Virginia City. Here one of the number, William Fairweather by name, washed a few pans of gravel, and was surprised to obtain about $2 worth of gold to the pan. The news soon spread, and numbers flocked to the place, which has since yielded $60,000,000 of gold, half of which was taken out during the first three years after the discovery. The next important placer diggings were found in 1864, at Last Chance Gulch, where Helena now stands, and at Silver Bow and German gulches, at the head of the Deer Lodge Valley. Subsequently mines of great richness were found at various other points, and the excitement upon the subject ran high.

The fame of the diggings caused a large immigration, and, with the honest and deserving gold hunters, there was also a rush of the vilest desperadoes from the mining camps of the Western States and Territories. This ruffianly element served as a nucleus around which the evil disposed gathered, and soon was organized a band of outlaws which became the terror of the country. These banditti included hotel keepers, express agents and other seemingly respectable people—Henry Plummer, the Sheriff of the principal county, being their leader. The roads of the Territory were infested by the ruffians, and it was not only unsafe, but almost certain death, to travel with money in one's possession. One writer affirms that "the community was in a state of blockade. No one supposed to have money could get out of the Territory alive. It was dangerous to cope with the gang, for it was very large and well organized, and so ramified throughout society, that no one knew whether his neighbor was or was not a member." The usual arms of a "road agent," writes Prof. Dimsdale, in his history of "The Vigilantes of Montana," "were a pair of revolvers, a double barreled shot gun of large bore, with the barrels cut down short, and to this was invariably added a knife or dagger. Thus armed, mounted on fleet, well trained horses, and disguised with blankets and masks, the robbers awaited their prey in ambush. When near enough, they sprang out on a keen run, with leveled shot guns, and usually gave the word 'Halt! throw up your hands, you ——— ———!' If this latter command were not instantly obeyed, that was the last of the offender; but in case he complied, as was usual, one or two of the ruffians sat on their horses, covering the party with their guns, which were loaded with buck shot, and one, dismounting, disarmed the victims, and made them throw their purses on the grass. This being done, a search for concealed property followed, after which the robbers rode away, reported the capture and divided the spoils."

At last the decent citizens organized a Vigilance Committee in self-defense. The confession of two of the gang put the lovers of law and order in possession of the names of the prominent ruffians, who were promptly arrested. Twenty-two of the miscreants were hung at various places, after the form of a trial, between December 21st, 1863, and January 25th, 1864, five having been executed together in Virginia City. This summary justice so stunned the remainder of the band that they decamped. From the discovery of the bodies of the victims, the confessions of the murderers before execution, and from information sent to the Vigilance Committee, it was found that certainly 102 people had been killed by the bandits in various places, and it was believed that scores of unfortunates had been murdered and buried, whose remains were never discovered. It was known that the missing persons had set out for various places with greater or less sums of money, and were never heard of again. After this wholesome justice had been meted to the murderers, law and order prevailed, the lawless element leaving the Territory, and the honest and enterprising remaining to develop the mining and other natural resources.

Beach—(174 miles west of Mandan).—This is the first station of the railroad in Montana. Beyond this fact the place is at present of no importance.

McClellan—(183 miles west of Mandan).—This is a bright new station, situated on Beaver Creek, a clear stream running over a gravelly bottom, and promises to develop into a pleasant little town. The soil is rich, the water pure and the point is a good one for cattle ranches.

Hodges and Allard are unimportant stations established on Glendive Creek. The Valley of Glendive Creek is noted for its attractive scenery.

The Yellowstone Valley.—The railroad follows up the Yellowstone Valley from Glendive to Livingston, a distance of 340 miles. In its characteristics the Yellowstone River more closely resembles the Ohio than any other American stream. Its waters, unlike those of the Missouri, are bright and clear, except when discolored by the freshets of its lower tributaries. The stream runs over a bed of gravel through permanent channels, and among thousands of beautiful islands, covered with heavy timber. It is navigable during a good stage of water for more than 250 miles, from its confluence with the Missouri at Fort Buford to a point above the mouth of the Big Horn River, by steam-boats of two or three hundred tons.

The Yellowstone has many tributaries along that part of its course which is traversed by the railroad, especially on its south bank. After leaving Glendive, the first important stream coming in from the south is the Powder River, so called by the Indians from its inky-black water, stained by the long course it runs through the alluvial soil, flanking the Black Hills and Big Horn Mountains. Here the valley of the Yellowstone broadens, and the country behind the bluffs is better and richer than before. On the north side of the Yellowstone, between Powder and Tongue rivers, several small streams come in which drain the divide between the Yellowstone and the Missouri. The next river of consequence on the south side is the Tongue, with a good but narrow valley, already well settled by farmers and herders. About thirty miles westward of the Tongue another affluent of considerable volume is the Rosebud, flowing from the south. Fifty-six miles beyond is the Big Horn River, the largest tributary of the Yellowstone, draining the whole eastern slope of the Rocky Mountains from the Yellowstone southward to the Platte. The next important stream is the Clark's Fork of the Yellowstone, which must not be confused with the other and more important Clark's Fork of the Columbia.

The Yellowstone wends from side to side of the valley, and along most of its course westward presents a very picturesque appearance. Bluffs of what are called "Bad Lands" enclose it, showing their precipitous faces against the stream, first on one side and then on the other, as the river winds from bluff to bluff, leaving always opposite the bluffs a considerable valley on either side of the stream. The width of the Yellowstone Valley throughout its entire length scarcely exceeds three miles; sometimes it narrows to not more than two miles, and again it widens to seven. At the heads of the lateral valleys are fine sites for stock ranches or grazing farms, the same luxuriant grasses covering the whole country. Clear, pure water is to be found every few miles in running streams and springs, along which are fringes of oak, ash, elm, box elder and cottonwood, with occasional pines and cedars in the ravines. Before reaching the Big Horn the valley becomes somewhat broader, and for many miles on the north side of the river, beginning at a point opposite Fort Keogh, are ranges of bluffs which finally recede in height and gradually disappear. Along this part of the river the rough, broken water-shed of the Musselshell, the Missouri and the Yellowstone, called the Bull Mountains, is drained by three small streams which have considerable valleys of fertile soil. The streams are Frozen Creek and the Big and Little Porcupine. The Yellowstone above the Big Horn runs through a comparatively narrow valley, which broadens only at a single point. The Clark's Fork Bottom lies in this part of the valley, on the north side of the Yellowstone, extending from the rocky bluffs east of the old settlement at Coulson, near the site of Billings, to the hills which put into the river from outlying spurs of the Rocky Mountains, some thirty-five miles westward.

The traveler, passing through the Yellowstone Valley, except during the months of May and June, when vegetation is vividly green, is apt to rebel against the withered look of the grass. Lowland and highland alike are clothed with a russet garment, which the heat of summer has spread over them. The mountains appear like colossal hay mows with the lush growth of bunch grass surging up their slopes, cured as it stands by the sun into the best of hay, upon which herds fatten all the year round. The valley has the same sere tone, and the fringe of dark pines on the brow of the hills does not relieve, but only serves to emphasize the prevailing tone of the landscape.

Yellowstone Division.

GLENDIVE TO BILLINGS.—DISTANCE, 225 MILES.

Glendive—(260 miles west of Mandan; population, 1,500.)—Glendive is the first place of any prominence in Montana that is reached by the railroad. It is the terminus of the Missouri Division and the beginning of the Yellowstone. The town is in latitude 47° 3' N., and longitude 104° 45' W., and lies 2,070 feet above the ocean level. Situated on the south bank of the Yellowstone, ninety miles from the junction of that stream with the Missouri, at Fort Buford, Dak., Glendive occupies a broad plain which slopes gently toward the river, and is sheltered by a range of curiously shaped clay buttes, distant about half a mile from the stream, and rising abruptly to a height of nearly 300 feet above its level. These buttes are not unlike those seen at the Bad Lands of the Little Missouri, only here the subterranean fires have not burned so fiercely as further east, and the river seems to have stopped the combustion, for across the water there is a large expanse of excellent soil. The site of the town was selected and laid out under the supervision of General Lewis Merrill, U. S. A., who adopted the name of Glendive for his projected city, in remembrance of Sir George Gore, an eccentric Irish nobleman, who spent the winter of 1856 in hunting buffalo in this vicinity, and who originally applied the designation to the creek.

Glendive was founded in 1881, and its progress has been steady and substantial. There are several flourishing business houses, the necessary stores and shops, public buildings, a bank, schools, church organizations, a very good hotel and a weekly newspaper. The soil in the neighborhood is a rich, sandy loam, and the gardens of the inhabitants yield fine vegetables. The surrounding country produces wheat, barley, corn, rye, oats and other crops. Wherever the land has been broken young trees have appeared spontaneously, and good water is obtained by digging wells to a

Eagle butte, near Glendive, Montana.

depth of from twenty to thirty feet. The place has already become important as a point of shipment to Eastern markets of cattle and sheep, as well as deer and buffalo hides, and, as it lies in the midst of an extensive grazing region, the stock interests are likely to become prominent.

The railroad company has built repair shops, round houses, a station and freight buildings at Glendive, the brick used in the construction of which was manufactured in the town.

This is an advantageous place for securing hunting outfits, ponies being comparatively cheap and provisions moderate. Game, usually abundant, embraces buffalo, elk, deer, antelope, bear, mountain sheep, timber and prairie wolves, jack rabbits and many varieties of birds. A tri-weekly line of stages runs between Glendive and Fort Buford, and also to Poplar Creek Indian Agency, on the Missouri River, ninety miles to the north.

The scenery just beyond Glendive is imposing. The railroad skirts the river, and bluffs tower several hundred feet above the track. *Eagle Cliff* is especially noticeable for its height, and the heavy engineering work which was necessary in constructing the railroad at this point.

At several stations between Glendive and Miles City an opportunity is afforded for passengers to see during the summer months stacks of thousands of buffalo hides, which the buffalo slaughterers have brought in from the ranges for shipment to market.

Iron Bluff—(10 miles from Glendive.)—This is the first station on the Yellowstone Division of the Northern Pacific Railroad. Large quantities of shell boulders are found in the vicinity. These consist chiefly of shells, which are mixed with small quantities of silica and alumina. The analysis shows seventy per cent. carbonate of lime, thirteen per cent. carbonate of magnesia, the remaining portion being silica, alumina and phosphate of lime. This shell conglomerate has been thoroughly experimented upon by Captain Maguire, of the U. S. Engineer Corps, who finds that it produces an excellent water lime, about equal in strength and quality to Louisville cement. There is a plentiful supply of this material in sight, but the extent of the deposit is not known.

Milton—(15 miles from Glendive)—is a small station with a section house for the railroad men.

Fallon—(29 miles from Glendive; population, 75)—is at the mouth of O'Fallon Creek. It is the depot for the beautiful and fertile valley running 100 miles south, which has attracted many ranch men and stock raisers.

Terry—(39 miles from Glendive ; population, 75.)—This is a bright, new place, named in honor of Brig.-Gen. Alfred H. Terry, U. S. A., and it is likely to become important. *Morgan*, ten miles beyond, is also a new station.

Ainslie—(59 miles from Glendive ; population, 100)—is the depot for the Powder River Valley region. The country around has a good soil, and there is sufficient rain-fall for raising all kinds of grain and vegetables. Several settlers have established themselves within a radius of three miles of the station, and most of the Government land has been taken up. Ten miles east of Ainslie, at the Powder River crossing, was fought a battle between the Indians and United States troops, and for several miles along the banks of the Yellowstone the graves of the soldiers who died of their wounds on their march up the river can be seen. *Dixon*, eleven miles further westward, is a new station, surrounded by a good country.

Miles City—(79 miles from Glendive ; population, 2,500)—is in longitude 104° W., latitude 49° 24′ N., and at an altitude of 2,350 above the ocean. It was founded in 1878, and derives its name from Brig.-Gen. Nelson A. Miles, U. S. A., whose campaigns against the hostile Indians in 1876-'77, resulted in the opening up of the Yellowstone Valley to settlement. Miles City is situated at the junction of the Tongue and Yellowstone rivers, and is surrounded by an excellent grazing country. Extensive and fertile valleys, tributary to this point, are rapidly filling up with a prosperous class of people. The waters of the Tongue River are diverted by a ditch from a point twelve miles above, and are brought down the valley for irrigating purposes and to supply the city with water. Miles City is the county-seat of Custer County, and its growth has been substantial. There are several large business houses, pleasant residences, two banks, one of which, with a capital of $250,000, transacts the greater part of the ranchers' business along the valley between Glendive and Billings, four hotels, and among them the Inter Ocean, situated near the railroad depot, which is the largest. There are also a court-house, a church, good schools, two daily and weekly newspapers, two public halls, two saw mills, a brewery, a good supply of trading stores and the usual shops. Friendly Indians do quite a business in buffalo robes and other furs. The streets are usually lively, with crowds of ranch men, travelers and Indians filling them. Dairy farms have been established in the neighborhood, and prove very successful. Game of all kinds is abundant within twenty miles of the place, the varieties consisting of buffalo, antelope, black and white tail deer, mountain sheep or big horn, elk, bears, and, in their season, ducks, geese and grouse. Hunting outfits are to be obtained here at reasonable rates. Good lumber is found near the city, and several large veins of coal have been opened in the vicinity. A tri-weekly stage runs from this town to Deadwood, Dak., and other points.

More than 15,000 head of young cattle were unloaded at Miles City during the spring of 1883, and driven thence to the adjacent ranges. The country on the north side of the Yellowstone is still mainly unoccupied, affording extensive ranges for cattle and sheep. The total amount of cattle on the ranges tributary to Miles City is estimated at 295,600.

Explorations of the Yellowstone.—The first recorded exploration of the Yellowstone Valley was that made by Captain William Clark, U. S. A., who was associated with Captain Meriwether Lewis, U. S. A., in the command of the famous Lewis and Clark expedition, fitted out in 1804, under authority of President Jefferson, to explore the region west of the Mississippi River and extending to the Pacific Coast. This vast territory, known as "the Louisiana purchase," and subsequently as the Province of Louisiana, was ceded to the United States by Napoleon Bonaparte, in 1803, for the nominal sum of $15,000,000. The heroic band of explorers, numbering only thirty-two men, set out from St. Louis, on the 14th of May, 1804, ascended the Missouri River a distance of 2,858 miles from its mouth, and striking across the Rocky Mountains and other ranges westward, reached the mouth of the Columbia River on the 7th of November, 1805. On the 23d of March, in the following year, the dauntless explorers entered upon their return journey, recrossing the Rocky Mountains on the 3d of July. The expedition now resolved itself into three parties, one of which followed the eastern base of the mountains northward to the mouth of the Marias River, where it united with the second party, commanded by Capt. Lewis, that had gone directly down the Missouri. The third detachment, under Captain Clark, pushed eastward until it struck the Yellowstone River, and then followed this stream 400 miles to its confluence with the Missouri, near which point the three parties again united. After an absence of nearly two years and a half the expedition arrived at St. Louis on the 23d of September, 1806, having lost only a single man by death. This was one of the most brilliant and successful explorations ever made. By its means a mass of accurate information respecting the country was gathered, the practical value of which has continued to the present day. The result of the expedition was at once to open up the newly acquired territory to the enterprise of the great fur companies, who established trading posts with the Indians at many points. Aside from the trappers, however, no whites settled in Montana until the breaking out of the gold excitement in 1862. Then, and even for many years afterwards, the settlements were confined to the extreme western portions of the Territory, which were the most accessible, the eastern half long remaining a wilderness, in absolute possession of the Indians.

Only since the year 1853, at which time the Government sent out an expedition, under command of the late General I. I. Stevens, to explore the region lying between the forty-seventh and forty-ninth parallels, with a view of reporting upon the feasibility of the northern route for a railroad from Lake Superior to Puget Sound, has the Yellowstone Valley been brought to public attention. Since the date named a number of expeditions, both Government and private, have passed through the valley from time to time, and their records of experience and adventure are of the highest interest. But it is not within the plan of this book even to outline the more important features of any of these exploring expeditions. The space at command will only admit of the narration of a few of the more important facts connected with the various conflicts between the Indians and the United States troops, of which this valley was the scene, between the years 1873 and 1877.

During the period in question the aborigines strove hard to keep possession of their favorite country. But civilization, repeating the history which has marked its progress in every land, was not to be kept back, and the fierce struggle for supremacy between the white race and the red man, resulted in the final disappearance of the latter from the Yellowstone Valley.

The railroad was finished to the Missouri River toward the close of 1872, but the actual surveys and locations for the roadway had been made as far west as the Powder River, 250 miles beyond. An escort of troops always accompanied the surveying parties, and minor engagements between these small detachments and the Indians were of common occurrence. During 1873 these attacks became so bold and frequent, that it was necessary to transfer an additional regiment of cavalry from the Military Department of the South for the purpose of holding the hostile red men in check, and a supply depot was established on Glendive Creek, where that stream empties into the Yellowstone.

A Fight with Indians at Tongue River.—In the summer of 1873 an army expedition, consisting of about 1,700 men, under the command of Major General D. S. Stanley, was sent out from Fort Rice, on the Missouri

River, to explore the Yellowstone Valley in the interest of the railroad. In due time the expedition reached the Yellowstone River, and marched for several days up that stream. The country eventually proved so rough and broken that in many places serious delays were encountered in finding a practicable route for the long and heavily laden wagon trains. These serious embarrassments were only overcome by sending out each morning, some distance in advance of the main column, two companies of the Seventh Cavalry, under command of the late General Custer, whose duty it was to seek and prepare a practicable road. In carrying out the plan, which already had been for some days followed successfully, General Custer left camp at five o'clock on the morning of the 4th of August, with a force of ninety-one men, guided by Bloody Knife, a young Arickaree warrior. Presently the watchful eyes of the scout discovered fresh signs of Indians. Halting long enough to inspect the trail and gather all the information possible, it was clear that a party of Indians had been prowling about the camp the previous night, and had gone away, traveling in the same direction as the detachment was marching. The discovery occasioned no surprise, as the presence of Indians had been expected for some days, and no change was made in the plan of the march, the hostile party numbering only nineteen and the detachment over ninety. But the thrilling episode which followed should be given in General Custer's own words, who himself graphically described it as follows :

"About ten o'clock we reached the crest of the high line of bluffs bordering the Yellowstone Valley, from which we obtained a fine view of the river and valley extending above and beyond us as far as the eye could reach. After halting long enough to take in the pleasure of the scene, and admire the beautiful valley, spread out like an exquisite carpet at our feet, we descended and directed our horses' heads toward a particularly attractive and inviting cluster of shade trees standing on the river bank, and distant from the crest of the bluffs nearly two miles. First allowing our thirsty horses to drink from the clear, crystal water of the Yellowstone, which ran murmuringly by in its long, tortuous course to the Missouri, we then picketed them out to graze. Precautionary and necessary measures having been attended to, looking to the security of our horses, the next and equally necessary step was to post half a dozen pickets on the open plain beyond, to give timely warning in the event of the approach of hostile Indians. This being done, the remainder of our party busied themselves in arranging each for his individual comfort, disposing themselves on the grass beneath the shade of the wide-spreading branches of the cottonwoods that grew close to the river bank. For myself, so oblivious was I to the prospect of immediate danger, that after selecting a most inviting spot for my noonday nap, and arranging my saddle and buckskin coat in the form of a comfortable pillow, I removed my boots, untied my cravat and opened my collar, preparatory to enjoying to the fullest extent the delight of the outdoor siesta. I did not omit, however, to place my trusty Remington rifle within easy grasp—more from habit, it must be confessed, than from any sense of danger. Near me, and stretched on the ground, sheltered by the shade of the same tree, was my brother, the colonel, divested of his hat, coat and boots; while close at hand, wrapped in deep slumber, lay the other three officers, Moylan, Calhoun and Varnum. Sleep had taken possession of us all—officers and men—excepting, of course, the watchful pickets, into whose keeping the safety, the lives, of our little detachment were for the time entrusted. How long we slept I scarcely know—perhaps an hour, when the cry of 'Indians! Indians!' quickly followed by the sharp, ringing crack of the pickets' carbines, aroused and brought us—officers, men and horses—to our feet. There was neither time nor occasion for questions to be asked or answered. Catching up my rifle, and without waiting to don hat or boots, I glanced through the grove of trees to the open plain or valley beyond, and saw a small party of Indians bearing down toward us as fast as their ponies could carry them.

"'Run to your horses, men! Run to your horses!' I fairly yelled, as I saw that the first move of the Indians was intended to stampede our animals, and leave us to be attended to afterwards. At the same time the pickets opened fire upon our disturbers, who had already emptied their rifles at us, as they advanced, as if boldly intending to ride us down. As yet we could see but half a dozen warriors, but those who were familiar with Indian stratagems knew full well that so small a party of savages, unsupported, would not venture to disturb, in open day, a force the size of ours. Quicker than I could pen this description, each trooper, with rifle in hand, rushed to secure his horse, and men and horses were soon withdrawn from the open plain and concealed behind the clump of trees, beneath whose shade we were but a few moments before quietly sleeping. The firing of the pickets, the latter having been reinforced by a score of their comrades, checked the advance of the Indians, and enabled us to saddle our horses and be prepared for whatever might be in store for us.

"A few moments found us in our saddles and sallying forth from the timber to try conclusions with the daring intruders. We could only see half a dozen Sioux warriors galloping up and down our front, boldly challenging us by their manner to attempt their capture or death. Of course, it was an easy matter to drive them away, but, as we advanced, it became noticeable that they retired, and when we halted or diminished our speed, they did likewise. It was apparent from the first that the Indians were resorting to stratagem to accomplish that which they could not do by an open, direct attack. Taking twenty troopers with me, headed by Captains Custer and Calhoun, and directing Moylan to keep within supporting distance with the remainder, I followed the retreating Sioux up the valley, but with no prospect of overtaking them, as they were mounted upon the fleetest of ponies. Thinking to tempt them within our grasp, I, being mounted on a Kentucky thoroughbred, in whose speed and endurance I had confidence, directed Captain Custer to allow me to approach the Indians, accompanied only by my orderly, who was also well mounted, at the same time to follow us cautiously at a distance of a couple of hundred yards. The wily redskins were not to be caught by any such artifice. They were perfectly willing that my orderly and myself should approach them, but at the same time they carefully watched the advance of the cavalry following me, and permitted no advantage. We had by this time almost arrived abreast of an immense tract of timber growing in the valley and extending to the water's edge, but distant from our resting place, from which we had been so rudely aroused, about two miles.

"The route taken by the Indians, and which they evidently intended us to follow, led past this timber, but not through it. When we had arrived almost opposite the nearest point, I signaled to the cavalry to halt, which was no sooner done than the Indians also came to a halt. I then made the sign to the latter for a parley, which was done simply by riding my horse in a circle. To this the savages only responded by looking on in silence for a few moments, then turning their ponies and moving off slowly, as if to say: 'Catch us if you can.' My suspicions were more than ever aroused, and I sent my orderly back to tell Captain Custer to keep a sharp eye upon the heavy bushes on our left, and scarcely 300 yards distant from where I sat on my horse. The orderly had delivered his message, and had almost rejoined me, when, judging from our halt that we intended to pursue no further, the real design and purpose of the savages were made evident. The small party in front had faced toward us, and were advancing as if to attack. I could scarcely credit the evidence of my eyes, but my astonishment had only begun when, turning to the wood on my left, I beheld bursting from their concealment between 300 and 400 Sioux warriors, mounted and caparisoned with all the flaming adornments of paint and feathers, which go to make up the Indian war costume. When I first obtained a glimpse of them—and a single glance was sufficient—they were dashing from the timber at full speed, yelling and whooping as only Indians can. At the same time they moved in perfect line, and with as seeming good order and alignment as the best drilled cavalry.

"To understand our relative positions, the reader has only to imagine a triangle whose sides are almost equal, their length in this particular instance being from 300 to 400 yards, the three angles being occupied by Captain Custer and his detachment, the Indians and myself. Whatever advantage there was in length of sides fell to my lot, and I lost no time in availing myself of it. Wheeling my horse suddenly around, and driving the spurs into his sides, I rode as only a man rides whose life is the prize, to reach Captain Custer and his men, not only in advance of the Indians, but before any of them could cut me off.

"Moylan, with his reserve, was still too far in the rear to render their assistance available in repelling the shock of the Indians' first attack. Realizing the great superiority of our enemies, not only in numbers, but in their ability to handle their arms and horses in a fight, and fearing they might dash through and disperse Captain Custer's small party of twenty men, and, having once broken the formation of the latter, despatch them in detail, I shouted at almost each bound of my horse: 'Dismount your men! Dismount your men!' but the distance which separated us, and the excitement of the occasion, prevented Captain Custer from hearing me. Fortunately, however, this was not the first time he had been called upon to contend against the sudden and unforeseen onslaught of savages, and, although failing to hear my suggestion, he realized instantly that the safety of his little band of troopers depended upon the adoption of prompt means of defense.

"Scarcely had the long line of splendidly mounted warriors rushed from their hiding place before Captain Custer's voice rang out sharp and clear: 'Prepare to fight on foot!' This order required three out of four troopers to leap from their saddles and take position on the ground, where, by more deliberate aim, and being freed from the management of their horses, a more effective resistance could be opposed to the rapidly approaching warriors. The fourth trooper in each group of 'fours' remained on his horse, holding the reins of the horses of his three comrades.

"Quicker than words can describe, the fifteen cavalry men, now on foot and acting as infantry, rushed forward a few paces in advance of the horses, deployed into open order, and, dropping on one or both knees in the low grass, waited with loaded carbines—with finger gently pressing the trigger—the approach of the Sioux, who rode boldly down as if apparently unconscious that the small group of troopers were on their front. 'Don't fire, men, till I give the word, and when you do fire, aim low,' was the quiet injunction given his men by their young commander, as he sat on his horse intently watching the advancing foe.

"Swiftly over the grassy plain leaped my noble steed, each bound bearing me nearer to both friends and foes. Had the race been confined to the Indians and myself, the closeness of the result would have satisfied an admirer even of the Derby. Nearer and nearer our paths approached each other, making it appear almost as if I were one of the line of warriors, as the latter bore down to accomplish the destruction of the little group of troopers in front. Swifter seem to fly our mettled steeds, the one to save, the other to destroy, until the common goal has almost been reached—a few more bounds, and friends and foes will be united, forming one contending mass.

"The victory was almost within the grasp of the redskins. It seemed that, but a moment more, and they would be trampling the kneeling troopers beneath the feet of their fleet-limbed ponies, when, 'Now, men, let them have it!' was the signal for a well directed volley, as fifteen cavalry carbines poured their contents into the ranks of the shrieking savages. Before the latter could recover from the surprise and confusion which followed, the carbines—thanks to the invention of breech-loaders—were almost instantly loaded, and a second carefully aimed discharge went whistling on its deadly errand. Several warriors were seen to reel in their saddles, and were only saved from falling by the quickly extended arms of their fellows. Ponies were tumbled over like butchered bullocks, their riders glad to find themselves escaping with less serious injuries. The effect of the rapid firing of the troopers, and the firm, determined stand, showing that they thought neither of flight nor surrender, was to compel the savages first to slacken their speed, then to lose their daring and confidence in their ability to trample down the little group of defenders in the front. Death to many of their number stared them in the face. Besides, if the small party of troopers in the front was able to oppose such plucky and destructive resistance to their attacks, what might not be expected should the main party under

Moylan, now swiftly approaching to the rescue, also take part in the struggle? But more quickly than my sluggish pen has been able to record the description of the scene, the battle line of the warriors exhibited signs of faltering, which soon degenerated into an absolute repulse. In a moment their attack was transformed into flight, in which each seemed only anxious to secure his individual safety. A triumphant cheer from the cavalrymen, as they sent a third installment of leaden messengers whistling about the ears of the fleeing redskins, served to spur both pony and rider to their utmost speed. Moylan, by this time, had reached the ground, and had united the entire force. The Indians, in the meantime, had plunged out of sight into the recesses of the jungle from which they first made their attack. We knew too well that their absence would be brief, and that they would resume the attack, but not in the manner of the first.

"We had inflicted no little loss upon them—dead and wounded ponies could be seen on the ground passed over by the Indians. The latter would not be satisfied without determined efforts to get revenge. Of this we were well aware.

"A moment's hurried consultation between the officers and myself, and we decided that, as we would be forced to act entirely upon the defensive against a vastly superior force, it would be better if we relieved ourselves, as far as possible, of the care of our horses, and take our chances in the fight which was yet to come, on foot. At the same time, we were then so far out on the open plain, and from the river bank, that the Indians could surround us. We must get nearer to the river, conceal our horses or shelter them from fire, then, with every available man, form a line or semi-circle, with our backs to the river, and defend ourselves till the arrival of the main body of the expedition, an event we could not expect for several hours. As if divining our intentions, and desiring to prevent their execution, the Indians now began their demonstrations looking to a renewal of the fight.

"Of course, it was easy to see what had been the original plan by which the Indians hoped to kill or capture our entire party. Stratagem was to play a prominent part in the quarrel. The few young warriors first sent to arouse us from our midday slumber came as a decoy to tempt us to pursue them beyond the ambush in which lay concealed the main body of the savages; the latter were to dash from their hiding place, intercept our retreat, and dispose of us after the most approved manner of barbarous warfare.

"The next move on our part was to fight our way back to the little clump of trees from which we had been so rudely startled. To do this, Captain Moylan, having united his force to that of Captain Custer, gave the order: 'Prepare to fight on foot!' This was quickly obeyed. Three-fourths of the fighting force were now on foot, armed with the carbines only. These were deployed in somewhat of a circular skirmish line, of which the horses formed the centre, the circle having a diameter of several hundred yards. In this order we made our way back to the timber, the Indians whooping, yelling and firing their rifles as they dashed madly by on their fleet war ponies. That the fire of their rifles should be effective under these circumstances could scarcely be expected. Neither could the most careful aim of the cavalrymen produce much better results. It forced the savages to keep at a respectful distance, however, and enabled us to make our retrograde movement. A few of our horses were shot by the Indians in this retrograde skirmish; none fatally, however. As we were falling back, contesting each foot of ground passed over, I heard a sudden, sharp cry of pain from one of the men in charge of the horses; the next moment I saw his arm hanging helplessly at his side, while a crimson current, flowing near his shoulder, told that the aim of the Indians had not been entirely in vain. The gallant fellow kept his seat in his saddle, however, and conducted the horses under his charge safely with the rest to the timber. Once concealed by the trees, and no longer requiring the horses to be moved, the number of horse holders was reduced, so as to allow but one trooper to eight horses, the entire remainder being required on the skirmish line. The redskins had followed us closely, step by step, to the timber, tempted in part by their great desire to obtain possession of our horses. If successful in this, they believed, no doubt, that flight on our part being no longer possible, we must be either killed or captured.

"Taking advantage of a natural terrace or embankment extending almost like a semi-circle in front of the little grove in which we had taken refuge, and at a distance of but a few hundred yards from the latter, I determined by driving the Indians beyond to adopt it as our breastwork or line of defense. This was soon accomplished, and we found ourselves deployed behind a natural parapet or bulwark, from which the troopers could deliver a carefully directed fire upon their enemies, and, at the same time, be protected largely from the bullets of the latter. The Indians made repeated and desperate efforts to dislodge us and force us to the level plateau. Every effort of this kind proved unavailing.

"Rather a remarkable instance of rifle shooting occurred in the early part of the contest. I was standing in a group of troopers, and with them was busily engaged firing at such of our enemies as exposed themselves. Bloody Knife was with us, his handsome face lighted up by the fire of battle, and the desire to avenge the many wrongs suffered by his people at the hands of the ruthless Sioux. All of us had had our attention drawn more than once to a Sioux warrior who, seeming more bold than his fellows, dashed repeatedly along the front of our lines, scarcely 200 yards distant, and, although the troopers had singled him out, he had thus far escaped untouched by their bullets. Encouraged by his success, perhaps, he concluded to taunt us again, and at the same time exhibit his own daring, by riding along the lines at full speed, but nearer than before. We saw him coming. Bloody Knife, with his Henry rifle poised gracefully in his hands, watched his coming, saying he intended to make this his enemy's last ride. He would send him to the happy hunting ground. I told the interpreter to tell Bloody Knife that at the moment the warrior reached a designated point directly opposite to us, he, Bloody Knife, should fire at the rider, and I, at the same instant, would fire at the pony.

"A smile of approval passed over the swarthy features of the friendly scout as he nodded assent. I held in my hand my well tried Remington. Resting on one knee and glancing along the barrel, at the same time seeing that Bloody Knife was also squatting low in the deep grass with rifle leveled, I awaited the approach of the warrior to the designated point. On he

came, brandishing his weapons and flaunting his shield in our faces, defying us to come out and fight like men. Swiftly sped the gallant little steed that bore him, scarcely needing the guiding rein. Nearer and nearer both horse and rider approached the fatal spot, when, sharp and clear, and so simultaneous as to sound as one, rang forth the report of the two rifles. The distance was less than 200 yards. The Indian was seen to throw up his arms and reel in his saddle, while the pony made one final leap, and both fell to the earth. A shout rose from the group of troopers, in which Bloody Knife and I joined. The same moment a few of the comrades of the fallen warrior rushed to his rescue, and, without dismounting from their ponies, scarcely pulling rein, clutched up the body, and the next moment disappeared from view.

"Foiled in their repeated attempts to dislodge us, the Indians withdrew to a point beyond the range of our rifles for the apparent purpose of devising a new plan of attack. Of this we soon became convinced. Hastily returning to a renewal of the struggle, we saw our adversaries arrange themselves in groups along our entire front. They were seen to dismount, and the quick eyes of Bloody Knife detected them making their way toward us by crawling through the grass. We were at a loss to comprehend their designs, as we could not believe they intended to attempt to storm our position on foot. We were not left long in doubt. Suddenly, and almost as if by magic, we beheld numerous small columns of smoke shooting up all along our front.

"Calling Bloody Knife and the interpreter to my side, I inquired the meaning of what we saw. 'They are setting fire to the long grass, and intend to burn us out,' was the scout's reply, at the same time keeping his eyes intently bent on the constantly increasing columns of smoke. His features wore a most solemn look; anxiety was plainly depicted there. Looking to him for suggestions and advice in this new phase of our danger, I saw his face gradually unbend and a scornful smile part his lips. 'The Great Spirit will not help our enemies,' was his muttered reply to my question. 'See,' he continued, 'the grass refuses to burn.' Casting my eyes along the line formed by the columns of smoke, I saw that Bloody Knife had spoken truly when he said: 'The grass refuses to burn.' This was easily accounted for. It was early in the month of August; the grass had not ripened or matured sufficiently to burn readily. A month later, and the flames would have swept us back to the river as if we had been surrounded by a growth of tinder. In a few moments the anxiety caused by the threatening of this new and terrible danger was dispelled. While the greatest activity was maintained in our front by our enemies, my attention was called to a single warrior, who, mounted on his pony, had deliberately, and, as I thought, rashly, passed around our left flank—our diminished numbers preventing us from extending our line close to the river—and was then in rear of our skirmishers, riding slowly along the crest of the low river bank with as apparent unconcern as if in the midst of his friends, instead of being almost in the power of his enemies. I imagined that his object was to get nearer to the grove in which our horses were concealed, and toward which he was moving slowly, to reconnoitre, and ascertain how much force we held in reserve. At the same time, as I can never see an Indian engaged in an unexplained act without conceiving treachery or stratagem to be at the bottom of it, I called to Lieut. Varnum, who commanded on the left, to take a few men and endeavor to cut the wily interloper off. This might have been accomplished but for the excessive zeal of some of Varnum's men, who acted with lack of caution, and enabled the Indian to discover their approach, and make his escape by a hurried gallop up the river. The men were at a loss even then to comprehend his strange manœuvre, but, after the fight had ended, and we obtained an opportunity to ride over and examine the ground, all was made clear, and we learned how narrowly we had escaped a most serious, if not fatal, disaster.

"The river bank in our rear was from twenty to thirty feet high. At its base, and along the water's edge, ran a narrow pebbly beach. The redskins had hit upon a novel, but, to us, most dangerous scheme, for capturing our horses, and, at the same time, throwing a large force of warriors directly in our rear. They had found a pathway beyond our rear, leading from the large tract of timber in which they were first concealed, through a cut or ravine in the river bank. By this they were enabled to reach the water's edge, from which point they could move down the river, following the pebbly beach referred to, the height of the river bank protecting them perfectly from our observation. Thus they would have placed themselves almost in the midst of our horses before we could have become aware of their designs. Had they been willing, as white men would have been, to assume greater risks, their success would have been assured. But they feared that we might discover their movements and catch them while strung out along the narrow beach, with no opportunity to escape. A few men on the bank could have shot down a vastly superior force. In this case the Indians had sent on this errand about one hundred warriors. After the discovery of this attack, and its failure, the battle languished for a while, and we were surprised to notice, not very long after, a general withdrawal from in front of our right, and a concentration of their forces opposite our left. The reason for this was soon made clear to us. Looking far to the right, and over the crest of the hills already described, we could see an immense cloud of dust rising and rapidly approaching. We could not be mistaken; we could not see the cause producing this dust, but there was not one of us who did not say to himself, 'Relief is at hand.' A few moments later a shout arose from the men. All eyes were turned to the bluffs in the distance, and there were to be seen, coming almost with the speed of the wind, four separate squadrons of Uncle Sam's best cavalry, with banners flying, horses' manes and tails floating on the breeze, and comrades spurring forward in generous emulation as to which squadron should land its colors first in the fight. It was a grand and welcome sight, but we waited not to enjoy it. Confident of support, and wearied from fighting on the defensive, now was our time to mount our steeds and force our enemies to seek safety in flight or to battle on more even terms. In a moment we were in our saddles and dashing after them. The only satisfaction we had was to drive at full speed for several miles a force outnumbering us five to one. In this pursuit we picked up a few ponies, which the Indians were compelled to abandon on account of wounds or exhaustion. Their wounded, of which there were quite a number, and their killed, as afterwards acknowledged by them when they returned to the agency to receive the provisions and fresh supplies of ammunition which a sentimental government, manipulated and directed by corrupt combinations, insists upon distributing annually, were sent to the rear before the flight of the main body. The number of Indians and ponies killed and wounded in this engagement, as shown by their subsequent admission, equaled that of half our entire force engaged.

"That night the forces of the expedition encamped on the battle ground, which was nearly opposite the mouth of Tongue River. My tent was pitched under the hill from which I had been so unceremoniously disturbed at the commencement of the fight; while under the wide-spreading branches of a neighboring cottonwood, guarded and watched over by sorrowing comrades, who kept up their lonely vigils through the night, lay the mangled bodies of two of our companions of the march, who, although not present nor participating in the fight, had fallen victims to the cruelty of our foes."

The victims in question were the veterinary surgeon and the sutler of the regiment, who, being civilians, were not required to keep the ranks, but could ride as they pleased. These men were murdered and scalped by the Indians, who surprised them as they lingered in the rear of the main column.

The engagement above described occurred near the mouth of the Tongue River; and, for a week afterwards, as the exploring party pursued its march, it entered upon a series of sharp skirmishes with a large force of Indians, who, however, were invariably repulsed, although the troops did not escape many severe casualties.

In 1874 and 1875 the Yellowstone Valley enjoyed comparative quiet, although there were hostile bands of Sioux roaming over the valleys of the Big Horn and Powder rivers, and the entire western frontier was ravaged by them. In June, 1875, a steam-boat expedition, consisting of seven officers and 100 men, commanded by Col. Forsyth, of Lieut.-Gen. Sheridan's staff, ascended the Yellowstone a distance of 430 miles, selecting sites for military posts at the mouth of the Tongue and Big Horn rivers, in order to better deal with the Indians. This expedition returned without encountering any hostile red men.

On February 21st, 1876, an expedition left Fort Ellis, near Bozeman, under command of Major Brisbin, numbering 221 officers and men, for the succor of a party of citizens, who were besieged by Indians at Fort Pease, near the confluence of the Big Horn with the Yellowstone. The original party consisted of forty-six men, who defended themselves desperately in a stockade until the relief column of troops arrived. Six persons were killed, eight wounded and thirteen escaped during the night, leaving only nineteen in the stockade, who were rescued by the troops.

Later, in 1876, the Government was compelled to send out a force against certain wild and hostile bands of Indians, who were roaming about Dakota and Montana, not only attacking settlers and immigrants, but also making war upon the Mandans and Arickarees, who were friendly to the whites. To this class belonged the notorious Sitting Bull, who was not a chief, but only a "head man," and whose immediate followers did not exceed thirty or forty lodges. Another disaffected chief was Crazy Horse, an Ogallala Sioux, who properly belonged to the Red Cloud Agency, and whose band comprised, perhaps, 120 lodges, numbering about 200 warriors. These bands had never accepted the agency system, and would not recognize the authority of the Government. They had been notified, however, by the Department of the Interior, that they must, before the 31st of January, 1876, retire to the reservations to which they were assigned, or take the alternative of being brought to subjection by the military power. Every effort, meanwhile, to pacify these bands proved unsuccessful. They refused to come into the agencies, settle down and be peaceable. A strong force of troops was, therefore, set in motion to subdue them. On the 1st of March, Col. J. J. Reynolds, with a force of 883 men, moved out from Fort Fetterman, on the North Platte River, in search of the hostiles, and, after marching through deep snow and suffering great hardship, reached the mouth of the Little Powder River on March 17th, at which point he attacked and defeated a large village of Sioux and Northern Cheyennes, under Crazy Horse, destroying 105 lodges and a great amount of ammunition and supplies, and capturing a large herd of animals. The troops, however, had suffered so much from the severity of the weather that they were compelled to return to Fort Fetterman to recuperate.

Operations were resumed by this force toward the end of the following May. On the 29th of that month a column of 1,000 men, under the command of Gen. Crook, again left Fort Fetterman, and on the 13th and 17th of June the Indians were discovered in large numbers on the Rosebud. Here a desperate fight took place, lasting several hours, resulting in the flight of the Indians after heavy losses. The casualties to the troops in this engagement were nine killed and twenty-one wounded. From the strength of the hostiles who attacked Gen. Crook's column, it now became apparent that not only Crazy Horse and his small band had to be fought, but also a large number of Indians who had reinforced them from the agencies along the Missouri, and from the Red Cloud and Spotted Tail Agencies, near the boundary line between Dakota and Nebraska. Under these circumstances,

Gen. Crook deemed it best to await reinforcements and supplies before proceeding further.

The Massacre of Custer's Command.—Simultaneously with Gen. Crook's operations, Gen. Terry had concentrated 400 infantry and 600 of the Seventh Cavalry, the latter under Gen. George A. Custer, at Fort Lincoln. With this force he left the fort on the 17th of May, and reached the mouth of the Powder River on the 7th of June, where a supply camp was established. From this point six troops of cavalry, under Major Reno, scouted up the Powder River to its forks, and across the country to the Rosebud, following down the last named stream to its mouth, definitely locating the Indians in force in the vicinity of the Little Big Horn River. During Major Reno's scout, the force under Gen. Terry moved up the south bank of the Yellowstone and formed a junction with a column consisting of six companies of infantry and four troops of cavalry, under Col. Gibbon, which had marched from Fort Ellis eastward, along the north bank of the Yellowstone, to a point opposite the Rosebud.

On June 21st, after a conference with Cols. Gibbon and Custer, Gen. Terry, who was in supreme command, communicated the following plan of operations: Gibbon's column was to cross the Yellowstone near the mouth of the Big Horn, march up this stream to the junction with the Little Big Horn, and thence up the latter, with the understanding that it would arrive at the last named point on June 26th. Custer, with the whole of the Seventh Cavalry, should proceed up the Rosebud until the direction of the Indian trail found by Reno should be ascertained. If this led to the Little Big Horn it should not be followed, but Custer should keep still further south before turning toward that river, in order to intercept the Indians should they attempt to slip between him and the mountains, and also in order, by a longer march, to give time for Col. Gibbon's column to come up. On the afternoon of June 22d Custer's column set out on its fatal march up the Rosebud, and on the morning of the 25th he and his immediate command were overwhelmed and pitilessly slaughtered by the Indians, who were concentrated in the valley of the Little Big Horn, to the number of over 2,500 fighting men. The harrowing details of the massacre are mainly a matter of conjecture. No officer or soldier who rode with their gallant leader into the valley of the Little Big Horn was spared to tell the tale of the disaster. The testimony of the field where the mutilated remains were found showed that a stubborn resistance had been offered by the troops, and that they had been beset by overpowering numbers. The bodies of 204 of the slain were buried on the battle-ground.

The Brilliant Work of Gen. Miles.—After this calamity had befallen the expedition, additional troops were sent to the scene of operations as rapidly as they could be gathered from distant posts, but too late to be of immediate use. The exultant Indians had already broken up their organization, and scattered far and wide as bands of marauders, placing themselves beyond the reach of punishment in a body. In the autumn most of the troops were withdrawn from Montana, leaving only a strong garrison, under the command of Gen. Nelson A. Miles, who was then Colonel of the Fifth Infantry, to occupy a cantonment at the mouth of the Tongue River (now Fort Keogh). Through the energy and bravery of this command, the Yellowstone Valley was soon entirely rid of the Indians. On October 10th a train of ninety-four wagons, with supplies, left Glendive for the cantonment at the mouth of Tongue River, and was beset the same night by Indians, seven or eight hundred strong, under Sitting Bull, who so crippled it that it was forced to turn back to Glendive for reinforcements. These obtained, it resumed its journey, the escort numbering eleven officers and 185 men, in the hope of getting the much-needed supplies to the garrison. On the 15th the Indians attacked once more, but were driven back at the point of the bayonet, while the wagons slowly advanced. In this way the train proceeded until the point was reached from which the return had been previously made. Here the Indians became more determined, firing the prairie, and compelling the wagons to advance through the flames. On the 16th of October an Indian runner brought in the following communication from Sitting Bull to Col. Otis, commanding the escort:

YELLOWSTONE.

I want to know what you are doing traveling on this road? You scare all the buffaloes away. I want to hunt in this place. I want you to turn back from here. If you don't, I will fight you again. I want you to leave what you have got here, and turn back from here. I am your friend.

SITTING BULL.

I mean all the rations you have got and some powder. Wish you would write as soon as you can.

Col. Otis replied to this cool request that he intended to take the train through, and would accommodate the Indians with a fight at any time.

The train moved on, the Indians surrounding it and keeping up firing at long range. Presently a flag of truce was sent in by Sitting Bull, who said that his men were hungry, tired of the war, anxious for peace, and wished Col. Otis to meet him in council outside the lines of the escort. This invitation was declined, but the Colonel said he would be glad to meet Sitting Bull inside the lines. The wary savage was afraid to do this, but sent three chiefs to represent him. Col. Otis told them he had no authority to treat with them, but that they could go to Tongue River and make their wishes known. After giving them a present of hard bread and bacon they were dismissed, and soon the entire body disappeared, leaving the train to pass on unmolested.

On the night of the 18th Col. Otis met Col. Miles, with his entire regiment, who had advanced to meet the train, being alarmed for its safety. Learning that Sitting Bull was in the vicinity, Col. Miles at once pursued him and overtook him at Cedar Creek. Here an unsatisfactory parley took place, Sitting Bull refusing peace except upon terms of his own making. The council broke up, the Indians taking position immediately for a fight. An engagement followed, the Indians being driven from the field and pursued forty-two miles to the south side of the Yellowstone. In their retreat they abandoned tons of dried meat, quantities of lodge poles, camp equipage and broken down cavalry horses. Five dead warriors were left on the field, besides those they were seen to carry away. The force of Col. Miles numbered 398 rifles, against opponents estimated at over 1,000. On October 27th over four hundred lodges, numbering about 2,000 men, women and children, surrendered to Col. Miles, and Sitting Bull, with his own small band, escaped northward. He was vigorously pursued, but the trail was obliterated by the snow, and the troops returned to the cantonment. Again, in December, a portion of the command, under Lieut. Baldwin, left their quarters in search of Sitting Bull, who was found and driven south of the Missouri, retreating to the Bad Lands. Less than two weeks afterwards the same command surprised Sitting Bull on the Redwater, capturing the camp and its contents, the Indians escaping with little besides what they had upon their persons, and scattering southward across the Yellowstone. Meanwhile, Col. Miles, with his main command, numbering 436 officers and men, had moved against the Sioux and Cheyennes, under Crazy Horse, in the valley of the Tongue River; and, after repeated engagements, lasting from the 1st of January to the 8th of the same month, over fields covered with ice and snow to a depth of from one to three feet, completely vanquished the hostiles and required them to surrender at the agencies. After the surrender of

Crazy Horse, the band of Sitting Bull, in order to escape further pursuit, retreated beyond the northern boundary, and took refuge upon British soil, where this troublesome Indian remained until the spring of 1883, at which time he returned to the United States, and was assigned to the Standing Rock Indian Agency, in Dakota. In May, 1877, Col. Miles led an expedition against a band of renegade Indians, under Lame Deer, that had broken away from those who had surrendered at Tongue River. This band was surprised near the Rosebud; and, while negotiations for a surrender were in progress, the Indians, either meditating or fearing treachery, began firing, and ended the parley. The fight was resumed, and the Indians were driven eight miles, fourteen having been killed, including the chiefs Lame Deer and Iron Star, and 450 horses and mules and the entire camp equipage fell into the hands of the troops. This band was afterwards pursued so hotly that it eventually surrendered at the Red Cloud and Spotted Tail agencies.

On the 18th of September, 1877, Col. Miles, having learned that the hostile Nez Percés, from Idaho, under Chief Joseph, pursued by Gens. Howard and Sturgis, were likely to reach the frontier before they could be overtaken, started out from his cantonment to intercept them. By a series of rapid marches on the flank of the hostiles, after traversing a distance of 267 miles, Col. Miles came up with the Nez Percé camp on the morning of September 30th at the Bear Paw Mountains, and compelled its surrender after a desperate resistance, with severe losses on both sides.

The troops under the command of Col. Miles, in their operations during the years 1876 and 1877, marched no less than 4,000 miles, captured 1,600 horses, ponies and mules, destroyed a large amount of camp equipage belonging to the hostiles, caused the surrender of numerous bands and cleared the country of upwards of seven thousand Indians. By this series of brilliant successes not less than 400 miles of the Yellowstone Valley were opened to settlement.

Current Ferries.—On the Yellowstone River, as well as on many other Western streams, a method of ferrying is in vogue which presents its peculiarities to Eastern eyes. The swift current is used as a motor for swinging a flat-bottomed ferry-boat over the river. An elevated wire cable is stretched from shore to shore. Pulleys, attached by stout ropes to either end of the boat, are geared to the cable. The craft is shoved off from the

brink at an angle oblique to the current and starts languidly, the pulleys moving spasmodically at first. Presently the full force of the tide is felt, and the pulleys spin along the cable, carrying the boat across at fine speed. Then, reaching the slacker water near the opposite shore, the pulleys resume the jerky progress on their cable track, and the boat grates upon the beach or puts her broad nose gently upon the strand precisely where it is wanted. The steering is done by means of a wheel, or, rather, windlass, used to taughten or slacken the pulley ropes, and so get the proper angle of resistance to the current. These ferry-boats scorn any suggestion of an ordinary rudder in the water. They are guided by the guy-ropes only. The ferry-men levy a dollar toll upon each horse and each wagon, which seems good pay for little labor. They lament, however, that the good old times are gone when five dollars was the ordinary tax for this service.

Fort Keogh—(81 miles from Glendive; population, 600)—is situated a mile and a half west of the Tongue River and two miles from Miles City, in a beautiful and fertile portion of the Yellowstone Valley. The fort was built in 1877 by Gen. N. A. Miles, and is the most important post in the Northwest, having a large garrison of infantry and cavalry, the numbers varying with the demands of other military stations on the frontier. Fort Keogh consists of a number of commodious barracks, hospital, school, chapels and other buildings, besides sixteen attractive cottage residences for officers and their families. The fort draws its supply of water from the Yellowstone, and feeds a pretty fountain in the square, about which the residences are arranged.

Lignite—(86 miles from Glendive; population, 30)—is a station established in the midst of a lignite coal district, and the locomotives take in their fuel at this point.

Horton and Hathaway—(distant respectively from Glendive 90 and 99 miles)—are new stations, established for the convenience of ranchmen in the fine grazing country southward.

Rosebud (110 miles from Glendive; population, 150)—is situated at the mouth of the Rosebud River. The extensive valley of this stream is admirably adapted to cattle raising, and its plains are dotted with settlements.

Forsyth—(126 miles from Glendive; population, 500).—The place is named in honor of Gen. James W. Forsyth, who was the first officer to land by steamer at the present site of the town, and for a long time it was known as Forsyth's Landing. It is situated in a delightful valley immediately on the banks of the Yellowstone River, and is surrounded by trees and immense bluffs rising abruptly on the south and west. Forsyth is the end of a freight train division, and the supply point for the settlers of the Rosebud bottom, on the south side, and the Big and Little Porcupine rivers, on the north side, of the Yellowstone. The town has five general merchandise stores, which do a large business. The Northern Pacific Railroad Company has a round-house and repair shops here. Stock yards have been laid out to meet the needs of large cattle shipments from this point. The land near Forsyth is under cultivation, yielding heavy crops of grain and vegetables.

Howard, Sanders and Myers—(distant respectively 134, 145 and 156 miles from Glendive)—are yet unimportant stations, serving to supply the needs of the settlers of the surrounding country.

Big Horn—(166 miles from Glendive, at the mouth of the Big Horn River)—is the diverging point for a country well adapted to stock raising. The valley of the Big Horn is fertile, and its enclosing hills are covered with excellent grazing. For a short distance above the mouth of the Big Horn the Yellowstone River is navigable by steamers of 300 tons.

The railroad crosses the turbulent waters of the Big Horn River, about two miles from the mouth of that stream, by a bridge 600 feet in length. Passing over the narrow intervening valley, it presently penetrates the bluffs which hem in the Yellowstone River, by means of a tunnel 1,100 feet long, and emerges into the comparatively small Yellowstone Valley beyond.

Custer—(172 miles from Glendive)—derives its name from a military post thirty miles south, named in honor of the late Gen. Custer. The place is half a mile distant from Junction City, the river landing, and eight miles from Etchetah. *Riverside*, 181 miles from Glendive, is an unimportant station.

Pompey's Pillar—(196 miles from Glendive)—is a mass of yellow sandstone, rising abruptly to a height of 400 feet, its base covering nearly

an acre of ground. About half way up, on the north side, is an inscription, of which the following is a miniature *fac-simile* :

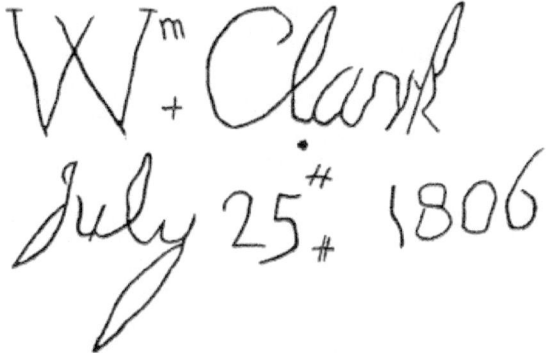

carved deeply in the rock by the explorer himself on his return journey across the continent. This inscription covers a space three feet long and eighteen inches high, and is surrounded by a border. It appears that Capt. Clark and his party were coming down the Yellowstone River in a boat, when they were overtaken by a storm which suddenly burst upon them. After it had cleared, they landed to examine a very remarkable rock, situated in an extensive bottom on the right, a short distance from the shore. "This rock," wrote the explorer, "is nearly 200 paces in circumference and about 200 feet high, accessible from the southeast only, the other sides consisting of perpendicular cliffs of a light-colored gritty stone. The soil on the summit is five or six feet deep, of a good quality, and covered with a short grass. The Indians have carved the figures of animals and other objects on the sides of the rock. From this height the eye ranges over a wide extent of variegated country. On the southwest are the Rocky Mountains, covered with snow. There is a low mountain about fifty miles distant, in a northwest direction, and at the distance of thirty-five miles the southern extremity of what are called the Little Wolf Mountains. The low grounds of the river extend nearly six miles to the southward, when they rise into plains reaching to the mountains, and are watered by a large creek, while at some distance below a range of highlands covered with pine stretches on both sides of the river in a direction north and south. The north side of the river for some distance is surrounded by jutty, romantic cliffs, succeeded by rugged hills, beyond which the plains are again open and extensive, and the whole country is enlivened by herds of buffalo, elk and wolves." After enjoying the prospect from this rock, to which Captain Clark gave the name of Pompey's Pillar, and carving his name and the date of his visit upon the stone, the explorer continued on his route. For the better protection of Captain Clark's name against vandals, who have already tried to cut their own insignificant designations within the border containing that of the heroic explorer, the railroad company has caused a screen to be placed over the relic for its protection.

The Crow Indian Reservation.—The entire southern shore of the Yellowstone River, from a point not far from Forsyth westward to the base of the Rocky Mountains, and extending south to Wyoming, was set apart by Congress, in 1868, as a reservation for the Crow Indians. This is one of the most fertile and best watered areas in Montana, including the valleys of all the large streams which flow into the Yellowstone above the Rosebud River. The reservation stretches along the Yellowstone for 200 miles, and has an average width of about 75 miles. Upon this territory, which is nearly as large as the State of Massachusetts, live not more than 3,000 Indians, who gather about the agency during winter, subsisting on Government beef and flour, and spend the summer in roaming about the country. They own 40,000 ponies, and are a very rich tribe from every point of view. The Crows have long been friendly to the whites, but they are far inferior to their old enemies, the Sioux, in intelligence, handicraft and bravery. In 1882 they re-ceded to the Government for a handsome consideration in cash a strip of the western end of their domain, about forty miles long by sixty wide, which embraces the Clark's Fork gold and silver mines, and it is only a question of time when the demands of the country for the release of all this valuable tract from its present possessors will be heard. Most of eastern Montana was originally claimed by the Crows, who at one time were a great and powerful nation. That the country was highly appreciated by these Indians is evidenced by the words of Arrapooish, a Crow chief, to the fur trader, Robert Campbell, as told in "Captain Bonneville's Adventures." by Washington Irving.

"The Crow country is a good country. The Great Spirit has put it exactly in the right place. When you are in it, you fare well; whenever you go out of it, whichever way you travel, you fare worse. If you go to the south, you have to wander over great, barren plains; the water is warm and bad, and you meet the fever and ague. To the north it is cold; the winters are long and bitter, with no grass; you cannot keep horses there,

Indian Camp on the line of the Northern Pacific Railroad.

but must travel with dogs. On the Columbia they are poor and dirty, paddle about in canoes, and eat fish. Their teeth are worn out; they are always taking fish-bones out of their mouths. To the east they live well, but they drink the muddy waters of the Missouri. A Crow's dog would not drink such water. About the forks of the Missouri is a fine country—good water, good grass and plenty of buffalo. In summer it is almost as good as the Crow country; but in winter it is cold, the grass is gone, and there is no salt weed for the horses. The Crow country is exactly in the right place. It has snowy mountains and sunny plains, all kinds of climate, and good things for every season. When the summer heats scorch the prairies, you can draw up under the mountains, where the air is sweet and cool, the grass fresh, and the bright streams come tumbling out of the snow banks. There you can hunt the elk, the deer and the antelope when their skins are fit for dressing; there you will find plenty of black bear and mountain sheep. In the autumn, when your horses are fat and strong from the mountain pastures, you can go down into the plains and hunt buffalo or trap beaver on the streams. And when winter comes on, you can take shelter in the woody bottoms along the rivers; there you will find buffalo meat for yourself and cottonwood bark for your horses. Or you may winter in the Wind River Valley, where there is salt weed in abundance. The Crow country is exactly in the right place. Everything good is to be found there. There is no country like the Crow country."

Huntley—(203 miles west of Glendive.)—This is a growing town, in the midst of a good stock-raising country. The land in its immediate vicinity is fast settling up, and it is a supply point for several large cattle ranges.

The Legend of Skull Butte.—The high and rugged elevation across the river to the left of the railroad, just before reaching Billings, is named Skull Butte. Tradition says that about seventy years ago several hundred lodges of Indians, belonging to the powerful Crow nation, were encamped on the river bottom, when small-pox broke out, and the ravages of the disease were so fearful that in a short time the tribe was decimated. To appease the anger of the Great Spirit, it was determined by the chief medicine man that forty young warriors should offer themselves as a sacrifice. Volunteers for this purpose were called for, and soon the allotted number of braves, who had recently passed through the ordeal of the "sun dance," and assumed the status of warriors, presented themselves. With much ceremony the preparation for the sacrifice was conducted, and after all the rites had been performed, the heroic band mounted their ponies, forded the river, ascended the steep heights opposite, and made themselves ready for their fate. It was determined that they and their horses should be blindfolded, and, rushing at full speed to the steep edge of the cliff, should plunge to the rocky strand hundreds of feet below. The word was given, and the forty braves, with tremendous shouts, urged their steeds to the brink of the cliff, and all went down to their destruction. For years afterwards, bleaching skulls and bones of men and horses were found around the base of Skull Butte.

The railroad crosses to the north side of the Yellowstone upon a substantial truss bridge, near the old settlement of Coulson, at the foot of Skull Butte.

TROUT FISHING ON THE BIG BOULDER.

Montana Division.

BILLINGS TO HELENA.—DISTANCE, 239 MILES.

Billings—(225 miles from Glendive; population, 2,200)—is named in honor of the Hon. Frederick Billings, late President of the Northern Pacific Railroad Company. It is situated at the foot of Clark's Fork Bottom, on a beautiful plain sloping down to the Yellowstone River, in the heart of the fertile and picturesque valley, and is the county-seat of the new County of Yellowstone. The town was founded in the spring of 1882, and contains over 400 buildings, with many more under construction. Among them are noticeable a substantial brick church edifice, capable of seating 300 persons, the gift of Mr. Billings, a handsome bank building, three hotels and several wholesale establishments. Good common schools are maintained, and a commodious school-house has been constructed. There are three newspapers, one daily and two weekly. This is the terminus of the Yellowstone Division and the beginning of the Montana Division of the railroad. The Company has built a substantial round-house, shops, etc., for the purposes of a division terminus, and gives employment to 150 men. The Headquarters Hotel, a fine structure, is near the track, and the Minnesota and Montana Land and Improvement Company, by whom the city was projected, are energetically at work in building up its interests in every direction, among the recent improvements being a hotel and passenger depot combined and a street railway. The Clark's Fork Bottom Ditch, thirty-nine miles long, terminating at Billings, is finished, and is designed to irrigate 100,000 acres of fertile soil. The valley, in which lies Clark's Fork Bottom, contains over 125,000 acres of excellent land, capable of producing all kinds of cereals and vegetables. Billings has tributary to it the Barker and Maginnis gold mining region, situated about 100 miles to the northward, and the Clark's Fork mines to the southeast. On one side, to the westward, are the great Musselshell and Judith valleys, and on the other, to the eastward, are the Little and Big Horn valleys. Large veins of coal of a good quality are found within a short distance of the town, and extensive beds of excellent coal exist in the Bull Mountains, about thirty miles distant. Active operations have begun for the construction of a branch road, which will run to these coal beds, and afterwards be extended through the Musselshell country to Fort Benton, 200 miles distant. Billings is a supply and trading point for a large extent of farming and grazing country, within a radius of over 100 miles. It also receives the trade of the Stinking Water District, Wyoming Territory, a large and prosperous tract of country. The town possesses extensive cattle yards, and is the principal cattle shipping point in Montana, great numbers of cattle being driven here for shipment from the Musselshell and Judith ranges. During the autumn of 1882 over 16,000 head were shipped. Billings is already a commercial centre, and its importance will grow with the settlement of the valley and of the Musselshell country, directly tributary to it. The Yellowstone River affords a fine water power for manufactories, there being a fall of eleven feet in a mile. Arrangements have been made for the erection of extensive slaughtering houses and the shipment of dressed beef in refrigerator cars. Large shipments of wool are made from here, and a good wool market is established. In May, 1883, an output from the Barker District of 60,000 pounds of silver bullion was shipped from Billings, and the reports from the Maginnis and Clark Fork mines are of a very encouraging nature.

Montana Stock and Sheep Raising.—Abundance of nutritious grasses, mildness of climate and markets easy of access, are a combination of advantages which render Montana famous as a cattle-raising region. Montana steers command the highest prices in the Chicago cattle mart, and the Northern Pacific Railroad, with over 700 miles of track within the Territory, affords ready transportation from the grazing fields to the East. All the better varieties of grass do as well in Montana as elsewhere, but the most valuable of the native grasses is the bunch grass. This grows most luxuriantly upon the high rolling plains, of which a large part of the surface of the Territory consists. It begins to renew itself in the early spring, before the ground is yet free from frost, rapidly attains its growth, is early cured, and stands as hay through the remainder of the year, until the succeeding spring. Throughout the winter months it perfectly retains its sweet and nutritious qualities. The manner of its growth is similar to that of the short, curly and quickly cured buffalo grass of the plains. It stands in detached clusters or bunches, between which are visible interstices of bare ground. Its clusters, however, are finer, denser, of much taller growth, and cover the ground more closely and compactly than the tufts of buffalo grass. A single acre of bunch grass is fully equal to three acres of average buffalo grass in the quantity it furnishes of actual sustenance for cattle. It is, moreover, a stronger nutriment than ordinary plains vegetation, being unexcelled by the best cultivated grasses, timothy hay or clover.

The railroad, except where the main line crosses the mountain ranges, follows a system of valleys, unsurpassed in their broad, beautiful and fertile surfaces, and extending across the Territory from east to west. These valleys are free to all for pasturage purposes. Over these great natural ranges the herds roam at will, being separated or "rounded up" by their owners only twice a year—in the spring to brand the calves, and in the fall to choose the fat steers for market. The principal cattle ranges of the Territory, aside from the great valley of the Yellowstone, are on the head waters of the Little Missouri, in the southeast; the valleys of the Powder, the Tongue, the Rosebud, the Big Horn (still in possession of the Crow tribe of Indians), and the Clark's Fork, which meet the Yellowstone region from the south; the great valley of the Sun River, the broad basin of the Judith, the magnificent valley of the Musselshell, all situated northward of the Yellowstone, and intermediate between the Bull, Belt, Big Snowy and Little Rocky ranges; the valleys of the east and west Gallatin, Madison and Jefferson rivers, adjacent to the eastern bases of the Rocky Mountains, and the intramontane country of the Clark's Fork of the Columbia, westward.

The customary way of managing a band of cattle in Montana is simply to brand them and turn them out upon the prairie. Under this careless management some steers are lost, which stray away or are stolen. A more careful system is to employ herders, one man for every 1,500 or 2,000 head of cattle, whose duty it is to ride about the outskirts of the range, follow any trails leading away, and drive the cattle back, seeking through neighboring herds, if there are any, for cattle that may have mistaken their companionship. At the spring round up, a few extra men have to be employed for several weeks. No human being dare go among the cattle on foot. If he did he would be gored or trampled to death at once. The animals are only accustomed to horsemen, of whom they are in wholesome terror; but the sight of a person on foot instantly causes a rush toward the strange appearance, and death is certain to him who fails to find a place of refuge. In starting a new herd, cows, bulls and yearlings are bought, but calves under one year old running with the herd are not counted

The average cost of raising a steer, not counting interest or capital invested, is from sixty cents to one dollar a year, so that a four-year-old steer raised from a calf and ready for market costs about $4. A herd consisting of yearlings, cows and bulls will have no steers ready for the market in less than two or three years. Taking into account the loss of interest on capital invested before returns are received, besides all expenses and ordinary losses, the average profit of stock raising in Montana during the last few years has been at least thirty per cent. per annum. Some well informed cattle men estimate it at forty or forty-five per cent.

A flock of sheep containing 1,000 head and upward, in good condition and free from disease, are procurable in western Montana for from $3 to $3.25 per head. They must be herded summer and winter in separate flocks of not more than 2,000 or 3,000 each, must be corraled every night and guarded against the depredations of dogs and wild animals. Hay must be provided to feed them while the ground is covered with snow, and sheds must be erected to protect them from severe storms. They must, moreover, be raised by themselves. Cattle and sheep cannot live together on the same range. The latter not only eat down the grass so closely that nothing is left for the cattle, but they also leave an odor which is very offensive to the others for at least two seasons afterward. But, notwithstanding that the cost of managing sheep is greater than that of handling cattle, the returns from sheep raising are quicker and larger. While a herd of young cattle begin to yield an income only at the expiration of three years, sheep yield a crop of wool the first summer after they are driven upon a range, and the increase of the band is much greater than that of cattle, being from seventy-five per cent. to 100 per cent. each year. The wool is of good quality, free from burrs, and brings a good price on the ranch, agents of Eastern houses being always on hand eager to buy it. The profits of sheep raising are generally estimated at a higher figure than those of cattle raising. The lowest calculation is based upon a net profit of from twenty-five to thirty-five per cent. on the whole investment, although occasionally larger returns reward the fortunate stockman.

There are few large bands of horses in Montana, but breeding these animals is beginning to receive attention. One of these bands is in the western upper Yellowstone Valley. A few years ago 209 mares were placed on this range, and they have increased to 1,500, worth an average of $75 each, while more than enough have been sold to pay all expenses. Breeders estimate that fifty brood mares and a draught stallion, costing in all $2,500, placed upon a stock ranch where the proprietor does his own herding, will

Driving Cattle from the Range to the Railroad.

in the course of five years be worth $10,000. Horses are more hardy than sheep or cattle, being better able to endure cold weather, and to "rustle" or paw through the snow that covers their pasturage. But they are so much more valuable than other species of stock, that most owners prefer to have their bands either fenced in or carefully herded. The best horse farms are those in small valleys, ten or twelve miles long, on whose sides the foot-hills extend up to high mountains. By fencing across the ends of such a valley the horses are prevented from straying.

The Cow Boys.—As the train passes through the Yellowstone Valley, it is no uncommon sight to see herds of sleek cattle contentedly grazing on the russet hills. Sometimes, also, droves of one or two thousand are noticed slowly advancing in a broad column from the direction of the distant mountains on their way to the railroad shipping stations. Such a drove is kept well in hand by a number of herders, picturesquely garbed in sombreros, gray shirts and buckskin breeches, each man being armed with rifle, revolvers, bowie knife and a rawhide whip, and mounted on a wiry pony. If the drove of cattle has made a march of several hundred miles from the range, it will be pioneered by a large band of ponies, carrying camp equipage and supplies, and serving as remounts for the cow boys. These latter are usually brawny, clear-eyed fellows, civil enough to answer questions, in spite of the fact that every fibre of both man and horse seems strained to its utmost tension in keeping the wilder and straying members of the drove within the bounds of the horned column. Looking at these active, hardy, well-built horsemen, who seem the perfection of physical vigor, it is hard to class them with the lawless ruffians who have sometimes terrorized a frontier town, and "cleaned it out" in the manner most approved by cow boys.

Grand Mountain Views.—In passing up the valley, westward of Billings, there is a prospect from the car windows which combines more striking features of beauty and grandeur than could hardly be found elsewhere nearer than Switzerland. Beyond the smiling valley and the winding, glistening river, to the westward and southward, rise white, gigantic masses of mountains. These snowy ranges are so lofty, and in some conditions of the atmosphere, so ethereal, that the surprise of an Eastern tourist, who had never seen high mountains before, was quite natural. Standing on the platform of a Pullman car his eye caught the white, gleaming bulwark on the western horizon. "Conductor, those clouds look very much like mountains," he said. "Clouds; what clouds?" replied the conductor, looking around the clear, blue sky. "Out there; just ahead of us." "Those are not clouds; they are the mountains at the head of the valley." "Good gracious!" exclaimed the traveler, who had got his conception of mountains from the Alleghanies or the Adirondacks. "Those white things way up in the sky mountains! Well, well, this is worth coming all the way from New York to see." Passing the unimportant station of *Carlton*, 18 miles west of Billings, the next stopping place is at

Park City (23 miles west of Billings; population, 250)—at the head of the Clark's Fork Bottom. Park City was settled in June, 1882. It is the centre of a large tract of agricultural land, the very last worthy of mention before the rough approaches to the Rocky Mountains are entered. Citadel Butte, three miles northeast of the town, commands from its summit, 400 feet above the plains, a fine view of the snowy peaks to the westward. Park City has good hotel accommodations, schools and stores representing various branches of trade. It is much resorted to by hunters and sportsmen, as deer, elk, antelope and small game are easily accessible, while the river and mountain streams abound in trout.

Reminiscences of Frontier Life.—Mr. Young, the postmaster at Park City, settled at Young's Point, two miles above the town, in 1877, and a high point of rocks to the left, where the Yellowstone makes a sharp bend southward, was named after him. Here he kept the stage ranch and post-office. At the time of his arrival the land was overrun with wild animals of every description, and the river teemed with trout, beaver and otter. Elk, with their calves, usually remained on the islands in the river during the summer months, and mountain sheep, with their young, haunted the high rocks on the opposite shore. Antelope and deer were so numerous that they were little hunted, and became tame and easily killed. Mr. Young relates that one winter, when there was an unusually heavy fall of snow, thousands of antelope were driven into the valley by scarcity of food, and would try and force an entrance into the corrals where tame cattle were feeding. They showed little signs of fear, hundreds being slaughtered by the settlers, who sometimes used nothing but an axe. Wild geese and ducks were plentiful, and their eggs were found in abundance on the islands. Gen. Sherman camped at Mr. Young's ranch in the autumn of 1877. The Nez Percé Indians were then causing trouble, and it was a great relief when the settlers were assured by him that they need have no fear. It seems, however, he did not know the danger, for a few days afterward couriers were sent out from Fort Ellis to notify settlers that the Indians were coming down the Clark's Fork, and would probably plunder or destroy everything in their way. On one of these days a son of Mr. Young was out hunting and saw a multitude of moving creatures coming down the Clark's Fork, but whether they were buffalo, antelope or horses he did not know. On returning that night and telling his father, the latter felt certain that the unknown objects were Indians. Next morning Mr. Young and his son went over the river on the Crow Reservation, and ascending a plateau, the father saw an Indian at a distance, the red man discovering him at the same moment. Mr. Young told his son, who had not yet been seen, to keep under cover of the hills, return home as fast as his horse could carry him, and do his best to protect the family. The Indian stood still until two others came up, when, after a short parley, one of them motioned with his hand for Mr. Young to go back, a signal which he was not long in obeying. The band crossed the Yellowstone near the mouth of the Clark's Fork that night. A terrible storm prevailed, and this was probably the reason that they did not come up the river to destroy the stage ranch property. Going down the river, they set fire to everything which was not too wet to burn, destroying hay stacks and buildings in their course. Three of the Indians went to a ranch where two men were eating breakfast, and, passing themselves off for friendly Crows, were invited to eat. They finished their meal, got up and shot the two men in the back, killing one instantly, the other only living long enough to tell what had happened. The main body proceeded further down the river to the next stage ranch, arriving just as the stage drove off. They captured the coach, took the mails and galloped the horses off across the prairie until the animals became completely fagged out, when they turned them loose, after which they ransacked the mails, scattering the letters broadcast over the plain. Gen. Howard was in hot pursuit down the Clark's Fork, but the marauders eluded him, turning northward and escaping beyond the Musselshell. Mr. Young returned home, after his narrow escape from the Indians on the Crow Reservation, and took his wife and son to a lumber camp four miles up the river. That night a Frenchman, who associated with the Indians, came with a band of thieving Crows and stole everything he had. The Indians rarely depredated upon Mr. Young's property, not even running off his horses. This immunity was probably due to the fact that the ground in the vicinity of his ranch, indicated by a large tree standing alone on the plain, and two large holes in the rock on the opposite side of the river (a short distance from and in full view of the cars), was medicine ground, where hostile tribes met to settle their difficulties, prepare their medicine, etc. In other words, this was a sacred place, and any white man living near who had the good will of the Indians was never molested. The wild beasts were almost as troublesome as the Indians. Mr. Young seldom brought in a haunch of venison without being followed by a pack of hungry wolves, howling and snapping at his horse's feet to his very dooryard.

Stillwater—(41 miles from Billings.)—At this point the railroad crosses again to the south side of the Yellowstone River. The Crow Indian Agency buildings are situated here, and the Indians assemble twice a year to receive supplies furnished by the Government. There are always a few Indians about the agency, but the larger part of the tribe is absent, except in winter, at the various hunting grounds.

Merrill, Reedpoint, Graycliff and *Dornix*, distant, respectively, from Billings, forty-nine, fifty-one, sixty-seven and eighty-one miles, are unimportant stations.

Piscatorial. The Yellowstone River, beyond its confluence with the Big Horn, flows with a strong current through a valley of varying breadth, and is fed by many beautiful mountain streams. Here trout are in abundance and give excellent sport. In passing over several hundred miles of the route in the autumn of 1882, before the railroad had got very far west of the Big Horn River, the writer had ample opportunity to indulge in the gentle pastime. The fish were plentiful at every place of bivouac. On one occasion the Big Boulder River, a broad, clear rushing stream, was reached half an hour in advance of the main party. Hastily putting a rod together, a cast of the fly was made, and the fish were found to be voracious. In forty minutes there were landed no less than seventeen beauties, several of which weighed two pounds each. This was done with due regard to sport. The tackle was delicate, and each fish had the chance to fight fairly for his liberty. Moreover, the fisherman was compelled to wade far out over the rough boulders in the river bed to reach his victims in their favorite haunt in a deep pool near the opposite shelving bank. This made it necessary to go back to the shore with each captive, after he was safe in the landing-net, the passage being made over slippery rocks in a strong current, and consuming much of the time. Compared with its size, what tremendous power a two-pound trout exhibits after it detects its mistake in snapping the deceptive fly! There is nothing in the way of sport more exhilarating than to subdue this wild outburst of vigor.

Springdale—(95 miles from Billings)—is the station for Hunter's Hot Springs.

Hunter's Springs.—These celebrated hot springs are situated eighteen miles east of Livingston, at the foot of the Crazy Mountains, on the north bank of the Yellowstone, one mile and three-quarters from the stream. They were noted for their wonderful healing virtues years before they became accessible by railroad; and, in fact, if the traditionary reports of the aborigines may be credited, have been famous among all the Northern tribes from time immemorial. All the Indians in friendly relations with the Crows—within whose country the springs were situated until their reservation lines were fixed by the Government—had for generations made pilgrimages to this natural sanitarium with their invalids, pitching their tepees around the fountains for the relief of their sick, while their sore-backed ponies were healed by washing them in the healing waters below. Of course, the curative properties of the springs were the last hope for those at a great distance, whose afflictions had baffled the skill of their ablest "medicine men." No better proof than this of the healing properties of the water could be afforded, as the savage tribes acquire all their knowledge of the treatment of diseases from the experience of ages, handed down from father to son. But there is abundant testimony, also, on the part of numbers of white men who have been restored to health by drinking and bathing in the water of these springs, that there was no superstition in the red man's faith in their remarkable curative powers. They are named Hunter's Springs, in recognition of the fact that Dr. A. J. Hunter was the first white man to visit them and discover their medicinal qualities. The doctor being in advance of the train with which he was traveling, and a mile north of its direct course—his object in making the detour being to capture an antelope or deer for dinner—was attracted to the springs by the cluster of Indian tepees which had been pitched around them. Eight or ten different tribes were represented in the concourse. He boldly rode into the promiscuous camp, and his friendly salutations were responded to in a spirit of equal friendliness. Being a physician, he perceived by the bright iron-stains upon the rocks, the strong sulphur fumes of the ascending vapors, and the white soda and magnesia coating of the vegetation growing out of the sedimentary deposits, the medicinal value of the waters. He reached the spot in the early part of July, 1864, his train being one of the first that entered the then newly discovered gold mines of Montana by way of the Big Horn Valley. Whoever may visit the now famous springs, and feast his eyes upon the beauties of the surrounding scenery, will not wonder that Dr. Hunter at once relinquished his bright hopes of winning fortune in the gold mines, and resolved that if any white man during his lifetime should become possessed of these healing fountains, he himself should be that man. Dr. Hunter now enjoys the fruition of the hopes that inspired him nineteen years ago. The clay all around the springs is a blue, adhesive, argillaceous formation, thickly studded with pyritic iron, some of the cubes shining with a gold-like lustre; and in close proximity to the hot water fountains there are copious springs, from which flow streams of pure water—as cold in the hottest weather as ordinary ice-water. But, valuable as his property is, Dr. Hunter has fully paid for it by the frequent risk of not only his own life, but of every member of his family. He moved his family to the springs in 1871, when marauding parties of Sioux Indians were constantly making raids throughout the country. Once his wife and two young daughters were alone when a war party dashed by their cabin—Bozeman, forty-five miles away, being the nearest white settlement; and on two other occasions the doctor and his little son exchanged shots with the red men. For five long years, or until the year following the massacre of Custer and his command, the proprietor of the springs and his family were constantly "in the midst of alarms."

Hunter's Springs are from 3,000 to 4,000 feet above the sea level, and from fifty to 100 above the Yellowstone River. Their temperature ranges from 148° to 168° Fahrenheit, and they discharge at least 2,000 gallons a minute—sufficient to accommodate all visitors, without the necessity of pumping. The water, hot or cold, is palatable, many who had used it while under treatment being regularly supplied with it by express, ordering it by the cask. The surrounding geological formations indicate that the springs have been flowing for many centuries. A chemical analysis shows sulphur to be the predominating constituent, but the water also contains magnesia, arsenic, iodine and lime.

The soil near Hunter's Springs is highly productive, being enriched with gypsum and other strong mineral fertilizers. Everything is produced in the gardens of this section that is cultivated in the States of Ohio, Indiana and Illinois. It is one of the best grazing localities in the Yellowstone Valley, the whole face of the country being heavily grassed.

Back in the bluffs, within easy walking distance of Hunter's Springs, there are still many antelope, while hares, ducks, geese and other small game abound in the vicinity. Deer are occasionally "jumped up" in the groves in the Yellowstone, near the Springs; and it is seldom that the

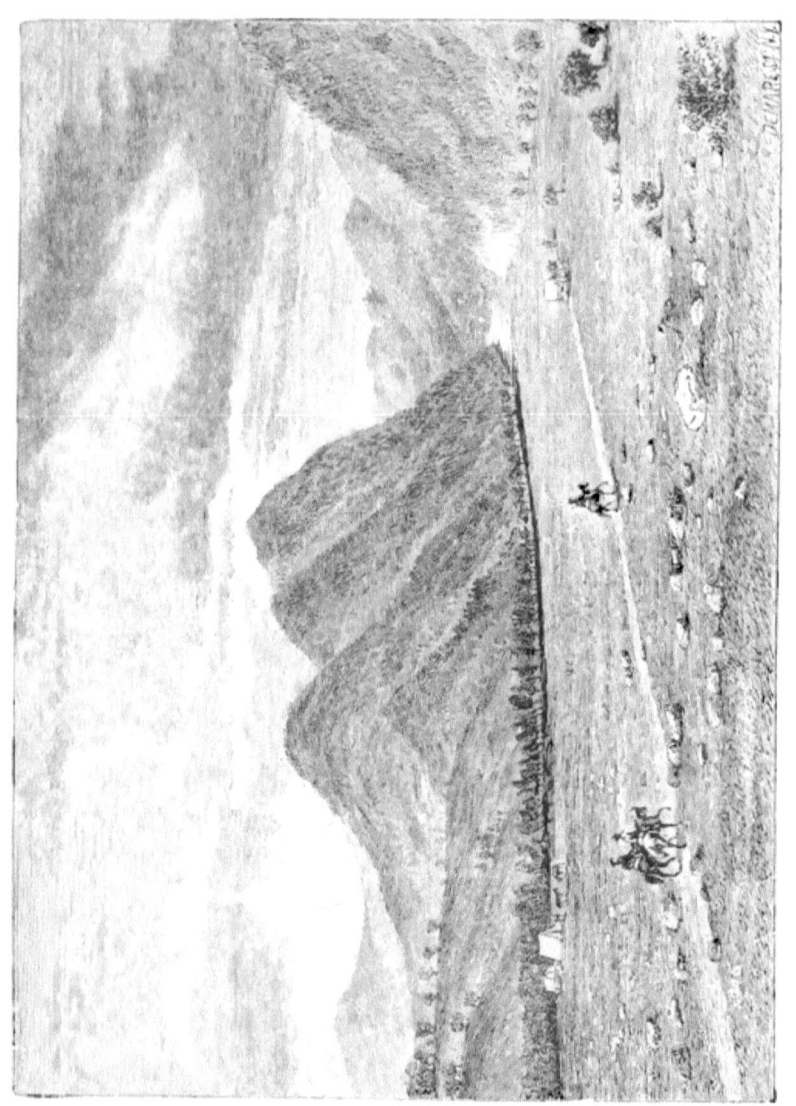

Gate of the Mountains, near Livingston.

sportsman walks far along its banks without having the opportunity to wing a goose or duck. Elk are numerous in the mountains a few miles out. Few rivers are more thronged with trout than the Yellowstone. The angler must be unskillful indeed, who fails to capture a handsome "string" in a couple of hours' fishing. The largest trout will weigh fully three pounds. Good coal has been found within two miles of Hunter's Springs; but the adjacent country has been only superficially prospected for minerals. Springdale Station is about three miles from this place, and there is telephonic communication between the two points. Mails arrive and depart daily. Hacks are at the station on the arrival of every train to take tourists and invalids to the springs. There are distinct bath-houses for the well and the sick, for male and female, and some of the tubs or tanks are large and deep enough for plunging and swimming. Visitors who prefer vapor baths are also accommodated; the medicated vapors coming up freshly from the steaming waters are regulated to any degree of temperature by cold-air jets.

Seven miles westward from Springdale is *Elton*, a new station, at which passenger trains do not stop.

Livingston—(115 miles from Billings; population, 1,500.)—This place is an important freight division and branch railroad terminus. Here the main line makes its third and last crossing of the Yellowstone River, leaving the valley, along which it has run a distance of 340 miles westward from Glendive, and passing through the Bozeman Tunnel, in the Belt Range of mountains, to the Gallatin Valley beyond. The river at this point makes an abrupt turn, flowing from its sources in the mountains far to the southward, through the world-renowned region of the Yellowstone National Park. Three miles from Livingston the high mountains of the Snow Range open their portals just wide enough to allow the river an outlet, and through the cañon thus cut by the stream the branch railroad to the Yellowstone National Park is laid. Livingston is situated on a broad, sloping plateau, on the left bank of the Yellowstone River, directly at the foot of the Belt Range. Less than a year ago the town was called into existence by the necessities of the railroad, and its growth has been necessarily rapid. Large engine houses, machine and repairing shops and other buildings for the use of the railroad are situated here, on a scale only second in magnitude to those at Brainerd. The railroad business alone is likely to give a permanent population of 1,500, and there are other reasons for believing that the growth and prosperity of the place are certain. Veins of fine bituminous coal, have been opened eight miles distant, and ledges of good limestone are in the immediate neighborhood. The Clark's Fork mines, rich in silver, lie directly south, and the surrounding hills are occupied by cattle ranches. All these are items which combine to render Livingston an important point. The bulk of travel to the Yellowstone National Park must also necessarily pass through Livingston, in view of which fact it is believed that a large business will be done in furnishing supplies to tourists. Hotel accommodation has been already provided, and various extensive business enterprises have been established. Livingston is one of the most convenient places from which to leave for the Crazy Mountains and the country adjoining them, which are the favorite breeding grounds of the elk. There is fine trout fishing in the vicinity of the town.

Yellowstone National Park.—The branch railroad to the Mammoth Hot Springs, leaving the main line of the Northern Pacific Railroad at Livingston, passes through the picturesque upper valley of the Yellowstone River to its terminus, fifty-seven miles distant. It does not come within the plan of this volume to describe the remarkable features of the Yellowstone National Park. It is believed that the convenience of the tourist has been best regarded by setting forth in detail the chief attractions of the Park in a separate book.*

Across the Belt Range.—After leaving Livingston the railroad runs for twelve miles from the Valley of the Yellowstone to the approach of the Bozeman Tunnel, on a grade of about 116 feet to the mile. The tunnel pierces the mountains a distance of 3,500 feet, at an elevation of 5,572 feet above the ocean. Some months before the completion of the work a short, steep grade track was laid over the summit of the pass for temporary use. It is far more agreeable to ride over the mountain than through it, and there are glorious views in every direction. The train runs down the western slope in the wild defile of Rock Cañon, passing out into the broad, fertile valley of the West Gallatin, at Elliston, near the military post of Fort Ellis, twenty-two miles from Livingston.

Bozeman—(141 miles from Billings; population, 2,000)—the county-seat of Gallatin County, is situated near the eastern end of the Gallatin Valley, at its narrowest point. North of the city the mountains are about

---

* Tourists are recommended to obtain a "Manual," descriptive of the Yellowstone National Park, profusely illustrated, which was prepared by the compiler of this volume, and is published by Messrs. G. P. Putnam's Sons, New York.

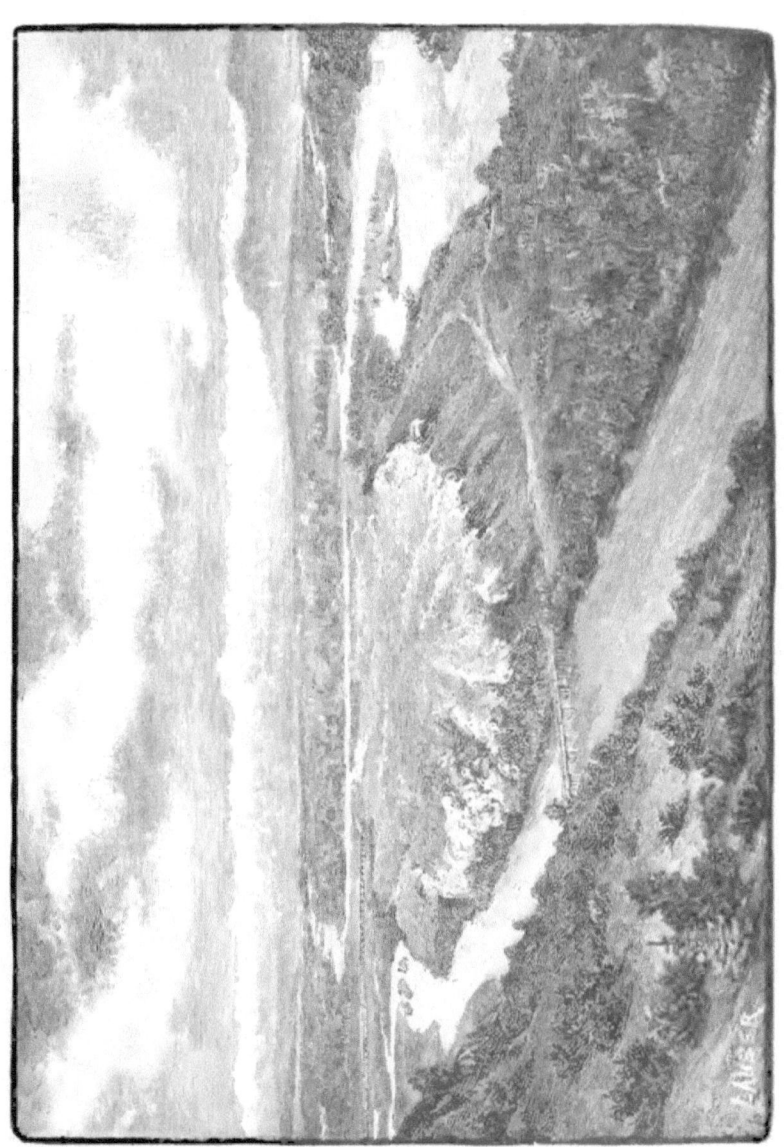

Three Forks of the Missouri—Gallatin, Madison, and Jefferson.

three miles distant, but the range suddenly diverges in the same direction, and afterwards the valley becomes twenty miles in width. Bozeman is the oldest established town on the line of the Northern Pacific Railroad in Montana, the town-site having been laid out in July, 1864. In August of that year a well-known frontiersman, John Bozeman, reached the place in charge of a party of emigrants, who were so impressed with the beauty and fertile soil of the valley that they determined to go no further. The town was named in honor of this pioneer, who was murdered three years afterwards by Indians in the Yellowstone Valley. For the first year or two the growth of Bozeman was slow. In 1865 a mill was put in operation, and two years afterwards Fort Ellis, situated two and a half miles east of the town, was established and garrisoned by three companies of United States troops. The gradual increase of population in the Gallatin Valley was soon evident, settlers coming in from the surrounding country and making Bozeman their trading centre. The city presents a very attractive appearance with its many substantial brick structures, among which are business blocks, churches, graded schools, and a fine court-house, while on every side appear handsome residences, and neat, cozy cottages. Large mercantile establishments form a prominent feature, many of which transact business to the amount of $100,000 a year. The city has also two flouring mills, three banking houses, two hotels, two planing mills and sash and door factories. Excellent brick is manufactured and used in the construction of the buildings. Lumber is abundant and cheap. There are two ably conducted newspapers of large circulation.

Bozeman owes much of her solidity to her agricultural resources. The Gallatin Valley is about thirty-one by twenty miles in extent, with a soil composed of a rich, dark vegetable mould. The cereal product of Gallatin County for the season of 1882 was 554,052 bushels of oats, 141,749 bushels of wheat and 7,560 bushels of barley, the market value of which was $574,282. Bozeman is also the head-quarters of some of the leading stock raisers and shippers of the Territory, and Gallatin County is largely developing its industry in this direction, as well as in dairy farming. The assessment lists of the county for the year 1882 show 10,158 horses, 37,700 cattle, 1,415 hogs, 18,115 sheep, aggregating in value $1,255,800. The development of the mineral resources of the county has also begun, and the results are encouraging. Locations have been made in the Clark's Fork, Emigrant Gulch, Bear Crevice, Mill Creek and Eight-Mile Creek districts, in the Upper Yellowstone country, and in the West Gallatin at Bridger, Sixteen-Mile Creek, Piney and other districts along the mountains skirting the Gallatin Valley, within sight of Bozeman. Work is being pushed on many old mines, and the immediate future promises a large production. Several veins of excellent bituminous coal are profitably worked a few miles east of the town, and outcroppings of silver, iron and copper are to be found on every side. Bozeman is a prominent outfitting point for the mines and cattle ranges, and for overland trips to distant points of interest. Daily lines of stages run to Fort Benton, Virginia City, Dillon, Butte City, and other important towns, and there is a tri-weekly stage to the Clark's Fork mines. The scenery surrounding Bozeman is very picturesque. Thermal Springs, said to contain medicinal properties, are within an hour's drive. Mystic Lake, twelve miles from the town, covers about eighty acres, and is a beautiful sheet of water. On the mountains around Mystic Lake, and in the vicinity of Bozeman, are forests of stately pines. Among the rivers in Gallatin County are the West Gallatin, Middle Fork and East Gallatin, the Madison, Yellowstone, Shield's River, Big and Little Timber, Sweet Grass, White Beaver, Kiser, Emigrant, Milk, Skull, Big and Little Boulder creeks, Stillwater, and many others of less importance. All these are stocked with trout and some other kinds of fish. Game, such as grizzlies, black bears, wapiti, deer and other animals, are abundant in the surrounding country.

The Three Forks, the Cañon and the Falls of the Missouri.—After leaving Bozeman, the road continues westward for thirty miles through the beautiful Gallatin Valley. There are many farms but no large villages on this part of the route. The junction of the Gallatin, the Madison and the Jefferson rivers is effected at Gallatin City, a hamlet whose existence is indicated by a flour mill, a few stores and a ranch. Here the three minor streams are merged into the Missouri, which flows northward 150 miles to Fort Benton, passing on its course through several grand cañons, whose walls rise from one thousand to two thousand feet above the river. At Fort Benton the Missouri turns eastward and southward, uniting with the Mississippi more than three thousand miles from its point of beginning in the Gallatin Valley. For all this distance the Missouri is navigable by steam-boats of at least two hundred tons burthen, except for eighteen miles at the Great Falls above Fort Benton. Many people visit these falls in small boats, which leave the Helena landing. Within a distance of ten miles above the Great Falls are ten others, varying in height from three to forty-seven feet, three of which, *viz.*, the Black Eagle, Rainbow and Crooked Falls, can scarcely be surpassed for perfection of form and graceful beauty.

For thirty miles beyond the Gallatin Valley, aside from the farming settlements which follow the banks of the Missouri, and are interspersed along the creeks which come down from the mountains, there is nothing on the immediate line of the railroad worthy of the name of a town. Across the Missouri River to the west, hidden away among the hills, a short distance left of the track, in the fertile Crow Creek Valley, is the town of *Radersburg*, the county-seat of Jefferson County. In the flush times of placer mining this was a very populous place, but now there are scarcely 200 inhabitants. There are many valuable quartz gold bearing lodes in the vicinity, and, doubtless, prosperity will return to Radersburg, as railroad facilities have reached its neighborhood.

Townsend—(207 miles from Billings; population, 350.)—This thriving town was laid out in the autumn of 1882 about one mile and a quarter eastward of the point at which the railroad crosses from the right to the left bank of the Missouri River, and received the name of Townsend. It is situated in the large and fertile valley of the Missouri, distant fifty-nine miles from Bozeman and thirty-nine from Helena, and was growing vigorously at the time these notes were written. Fifty or sixty houses were in existence, among them an hotel, and a flouring mill was in course of construction. Townsend will doubtless be the distributing point for a large and fruitful section of the Missouri Valley. It is the nearest station to the celebrated White Sulphur Springs, at the head of Smith River, which are a favorite summer resort, and also for many remarkable places of picturesque beauty in the mountains.

Across the Missouri Valley, in a northeasterly direction, a series of deep gorges or cañons has been cut by the waters in the faces of the precipitous mountains. Crowning the summits of the first range skirting the valley is a gigantic ledge of lime rock. This ledge has been thrown up in places to a great height with almost vertical sides, which are partly smooth, partly seamed and gashed by ages of storms, and sometimes cut through from top to bottom by the streams, forming narrow gorges of fantastic shapes. Avalanche Cañon is of great note, from its wild beauty and extensive and rich placer mines. This cañon received its name from the frequency of avalanches or snow slides, which rush down its almost perpendicular sides in winter, sometimes completely filling the gorge. Hell Gate Cañon, about two miles westwardly, while having a peculiarly suggestive name, amply merits the appellation. Perhaps in no other accessible spot in Montana is there as much rugged beauty in so small a place. The cañon forms the tortuous passage of a silvery stream through a series of gates cut in very high walls. These gates are so narrow that a man can span their width with extended arms. The walls are only a few feet in thickness, but of a surprising height. On each side of the main gorge are smooth fissures, called Devil's Slides, and every nook is filled with bright mosses and lichens.

Two miles beyond the crossing of the Missouri River the railroad passes through the mining town of *Bedford*, which was established in 1864. Much placer mining is still carried on in the vicinity, and the country for miles and miles around is covered with enormous piles of stones and earth, the *débris* of the diggings. The road follows down the valley of the Missouri, past one or two old and thrifty settlements, to Helena, twenty miles beyond.

SITTING BULL.

Rocky Mountain Division.

HELENA TO HERON.—DISTANCE, 274 MILES.

Helena—(239 miles from Billings; population, 7,000.)—This is the terminus of the Montana Division, and the beginning of the Rocky Mountain Division of the railroad. Helena, the capital of Montana, is pleasantly situated at the eastern foot of the main chain of the Rocky Mountains, in latitude 46° 30′ N., and longitude 112° 4′ west of Greenwich, on both sides of the famous Last Chance Gulch, from which at least ten millions of dollars worth of nuggets and gold dust have been taken, and which still yields annually a considerable amount of the precious metal. So large was the influx of miners at this point in 1864, that the United States Government felt compelled to establish a post-office for their accommodation. Until then the camp had been known as "Crab Town," but a meeting was called for the purpose of selecting a better name, and the majority of those assembled decided upon christening it after Helen of Troy. The city is the commercial and financial centre of the Territory, and the converging point of all the stage, express and telegraph lines. It contains a public and also a Territorial library, a classical school, a graded public school, with fine school-houses in different parts of the city; a theatre, with seating capacity for 1,200 people; six churches, Episcopal, Roman Catholic, Baptist, Methodist North, Methodist South, and Presbyterian; the United States Assay Office; Societies of Odd Fellows and Knights of Pythias. There are four National Banks, with over $3,000,000 on deposit, a Board of Trade, a well organized Fire Department, equipped with three engines and electric fire alarms; German singing and Turner societies; an art club, several good hotels, imposing business blocks, and many beautiful private residences. Pure, cold spring water is abundantly supplied from the surrounding mountains, and the streets are illuminated by electric lights. Foundries, saw, grist and planing mills, wagon factories and other industries are situated near the city, and there is telephone communication within the city and also with the mining camps within a radius of fifty miles. Perfect drainage is insured by the fact that Helena lies on a long slope, at the foot of which spreads out the beautiful Prickly Pear Valley, twenty-five miles long by twelve wide, oval in shape, and thickly studded with farms, the soil of which has produced 100 bushels of oats to the acre.

Helena is surrounded by mountains, rising one above the other, until the more distant are lost among the clouds, forming a view of striking beauty and grandeur, which is visible from every part of the city. To the south and west these mountains recede in long, picturesque, timbered ridges to the main range of the continental divide. The Missouri River is only twelve miles distant, and eighteen miles north of the city begins the famous cañon of the Missouri River, named by Lewis and Clark's expedition in 1805 "The Gates of the Rocky Mountains." Here the river has forced its way through a spur of the Belt Mountains, forming cliffs, frequently vertical, from 500 to 1,500 feet high, which rise from the water's edge for a distance of twelve miles. Near the lower end of this wonderful cañon, in plain view of Helena, thirty miles distant, is the jagged peak called by the Indians "The Bear's Tooth," rising abruptly from the river to a height of 2,500 feet, and almost hanging over the head of the voyager as he floats down the stream.

To the left of this curious object a few miles, and breaking through the same range of stratified mountains, is the cañon of Little Prickly Pear Creek, a magnificent chasm some fifteen miles long, with an endless variety of views of lofty cliffs crowned with pines and romantic dells and gorges, where the cottonwood and the alder hang over deep, shady pools, in which hundreds of trout await their destiny in the shape of the man with a bamboo rod and book of flies. This cañon, with hotel accommodation at each end of it, is accessible by carriages, as well as by a stage line of "palace jerkies," which passes through it three times a week for Fort Benton. "The Gates of the Rocky Mountains" are reached either by carriages to the upper end of the cañon or by boat through the cañon itself.

Among the other attractions of Helena are the Hot Springs, situated in a romantic glen, four miles west of the city, which are much resorted to by persons afflicted with rheumatism. The temperature of the water as it bubbles up from the earth varies from 110° to 140° Fahrenheit.

About four miles southwest, at the head of Grizzly Gulch, is a group of rich quartz mines, and also some placer diggings, both of which have been extensively worked. There are here many quartz mills, and the

The Gates of the Rocky Mountains, Missouri River, near Helena, Mont.

drive from the city is through pleasant mountain scenery. Twenty miles to the northwest, over a fine road, are several mining districts, in which are some of the richest gold and silver mines in the Territory. These are worked by a number of large quartz mills, around which have grown up picturesque mountain villages that will amply repay the trouble of a visit. Twenty-eight miles to the southeast are the mining towns of Jefferson City, Wickes and Clancy, in the vicinity of which are a great number of rich and extensive silver mines, which are worked by the smelting process, and give employment to many men and teams. Fifteen miles west, at the head of Ten-Mile Creek, is a rich belt of silver mines. Thus Helena is surrounded on all sides by rich mining districts, which are in a great measure tributary to her.

Making Pemmican and Buffalo Robes.—The great "fall hunt" of the Indians is the most important epoch of their year. In September the Flatheads, Pend d'Oreille, Spokane and Nez Percé tribes leave their reservations west of the Rocky Mountains, accompanied by their women and children, and wander over the eastern plains to hunt buffalo and steal cattle and horses. They are all mounted, dressed in full Indian costume, and their scanty effects and buffalo skin lodges are packed upon ponies. The squaws, with due regard to their own convenience and to the comfort of their horses, always ride astraddle. Even children, not more than two or three years of age, are lashed upon the backs of animals, which are turned loose with the rest of the band, and these infant riders seldom meet with an accident. The Western Indians lead a life of much hardship during the progress of the hunt, and not infrequently come in conflict with the Sioux, Crows, Blackfeet, Crees and other tribes of the plains, who regard the strangers as interlopers and poachers. These encounters usually result in the loss of a few lives, as well as active horse stealing on a grand scale. The men confine their energies to hunting, fighting and stampeding the horses of their opponents. After the buffaloes have been killed, the harder work of skinning the carcasses and curing the meat is performed by the women. A slit is first made along the length of the animal's back, and the hide is quickly stripped off with the least possible damage or mutilation. The subsequent processes are thus described by Col. Richard I. Dodge, in his interesting volume on the "Plains of the Great West."

"The skins are spread, flesh side upward, on a level piece of ground; small slits are cut in the edges of each, and they are tightly stretched and fastened down by wooden pegs through the slits into the ground. The meat is cut into thin flakes and placed on the drying scaffolds or poles. All this work is done in an incredibly short time. Another surround is then made by the hunters, and so on until the winter's supply is obtained. The hunt being over, or, in the intervals, if game is scarce, the women proceed to 'gather the crop.' Old *parflèches* (made of buffalo skins) are brushed up and new ones made. The now thoroughly dried meat is pounded to powder between two stones and packed in these receptacles. Melted tallow is poured over the whole, which is kept warm until the mass is thoroughly saturated. When cold the *parflèches* are closed and tightly tied up. The contents so prepared will keep in good condition for several years. The dressing of skins is the next work. No tannin is used, consequently no leather is made. The thickest hides are selected for shields, *parflèches*, etc. The hair is taken off by soaking the skins in water in which is mixed wood ashes, lime or some natural alkali. The skin is then cut into the required shape and put on a form while green. When it becomes dry it retains its shape and is almost as hard as iron.

"Making a robe is a much more difficult process. The skin in its natural condition is much too thick for use, being unwieldy and lacking pliability. This thickness must be reduced at least half, and the skin at the same time made soft. When the stretched skin has become dry and hard from action of the sun the woman goes to work upon it with a small iron instrument shaped somewhat like a carpenter's adze. It has a short handle of wood or elk horn, tied on with raw hide, and can be used with one hand. These tools are heirlooms in families and are greatly prized, more especially those with elk-horn handles. With this she chips at the hard skin, cutting off a thin shaving at each blow. The skill of this process is in so directing and tempering the blows as to cut the skin, yet not cut through it, and in finally obtaining a perfectly smooth and even inner surface and uniform thickness. To render the skin soft and pliable every little while the chipping is stopped, and the chipped surface smeared with fat and brains of buffalo, which are thoroughly rubbed in with a smooth stone. It is a long and tedious process, and none but an Indian would go through it. Hides for making lodges have the hair taken off, are reduced in thickness and made pliable. Deer, antelope or other thin skins are beautifully prepared for clothing, the hair being always removed. Thus there are four different processes in the preparation of skins, each admirably adapted to the use to which the prepared skin is to be put."

Across the Main Divide.—About nineteen miles from Helena the main range of the Rocky Mountains is crossed by the railroad at the Mullan Pass, so named after Lieut. John Mullan, U. S. A., who in 1867 built a wagon road from Fort Benton, Mont., to Fort Walla Walla, W. T., thus bringing these distant military posts into direct communication. Here there is a tunnel 3,850 feet in length, and 5,547 feet above the level of the ocean, lower by more than 2,500 feet than the highest elevation of the Union Pacific Railroad, and 1,200 feet below the highest elevation on the line of the Central Pacific. The tunnel is not yet finished. Meanwhile, as at the tunnel near Bozeman, a steep grade track has been laid over the brow of the mountain for the accommodation of the lighter traffic. Perhaps this temporary track will be permanently maintained, in order that summer tourists may always enjoy the inspiring scenery from the small, level plateau on the crest of the rocky heights, where the waters of the Atlantic slope are divided from those of the Pacific. The route from Helena to the Mullan Pass is through the charming valley of the Prickly Pear, across Ten Mile Creek, and up, past heavy growths of pine and spruce and masses of broken boulders, the narrow basin of Seven Mile Creek to the eastern portal of the tunnel. The scene from above reveals one of the most picturesque regions in Montana, in which mountain and valley, forest and stream are all conspicuous features. Describing this region in a recent letter, Mr. E. V. Smalley wrote:

"Approached from the east, the Rocky Mountains seem well to deserve their name. Gigantic cliffs and buttresses of granite appear to bar the way, and to forbid the traveler's further progress. There are depressions in the range, however, where ravines run up the slopes, and torrents come leaping down, fed by melting snows. Over one of these depressions Lieut. John Mullan built a wagon road a score of years ago, to serve the needs of army transportation between the head of navigation at the Great Falls of the Missouri and the posts in Oregon. Mullan's wisdom in selecting the pass, which bears his name, was endorsed when the railroad engineers found it to be the most favorable on the Northern Pacific line. The road is carried up ravines and across the face of foot-hills to a steep wall, where it dives into the mountain side, runs under the crest of the Divide through a tunnel three-quarters of a mile long, and comes out upon smiling green and flowery meadows, to follow a clear trout stream down to a river whose waters seek the mighty Columbia. The contrast between the western and eastern sides of the Main Divide of the Rockies is remarkable. On the eastern slope the landscapes are magnificently savage and sombre; on the western slope they have a pleasant, pastoral beauty, and one might think himself in the hill country of western Pennsylvania instead of high upon the side of the great water-shed of the continent. The forest tracts look like groves planted by a landscape gardener in some stately park, and the grassy slopes and valleys, covered with blue and yellow flowers, and traversed by swift, clear brooks, add to the pleasure-ground appearance of the country. What a glorious place this would be for summer camping, trout fishing and shooting, is the thought of every traveler as he descends from the summit, with his hands full of flowers picked close to a snow-bank. Snow Shoe Mountain rises just in front, across a lovely verdant valley. Powell's Peak, a massive, white pyramid, cuts the clear sky with its sharp outlines on the further horizon, and a cool breeze blows straight from the Pacific Ocean."

Passing down the western slope, the descent is made to the valley of the Little Blackfoot River. This valley is open and well grassed, with cottonwood on the stream and pine on the slopes of the hills. The river received its name from the Blackfeet Indians, who often passed down the valley to make their raids upon the settlers in Deer Lodge and Missoula counties. Their last exploit of this character was in April, 1864, when they stole 180 horses in one night from the Deer Lodge Valley. They were pursued by five men, and overtaken after a chase of 120 miles. There were only eight in the party, and when the pursuers came up, the thieves mounted eight of the best horses, and, abandoning the others, escaped.

There is good ruffled grouse shooting in the valley, and also a great many blue grouse in the neighboring cañons. In October black-tailed deer are plentiful, and elk are also found in the mountains. Even a few bison manage to conceal themselves in the mountain fastnesses. Bear—black, grizzly and cinnamon—are numerous, and it is not at all uncommon for them to hunt the hunters, an instance of which appears in the following incident:

An Adventure with a Grizzly.—William Roe, an old resident of Helena, while hunting alone on the mountains northwest of town, saw a large grizzly bear on a little knoll below him, about 125 yards distant, across a hollow filled with dead and much fallen timber. He was armed with a forty-five calibre, seventy-five grain Winchester rifle, which had twelve cartridges in the magazine. Taking good aim, he fired; the bear, which had just discovered him, fell as though mortally wounded; but, after tearing at the wound for a few seconds, it rose to its feet and made for him. It had advanced only a few paces when he shot once more, again knocking the animal down, but, as before, it immediately rose and rushed for him. He again knocked it down, and continued to do so with the same result, until, as it reared up to cross a huge fallen tree that lay in its path, he shot it for the seventh time full in the breast, and apparently with fatal effect, for it fell over backward and lay with its feet in the air for a few moments. Thinking it dead, Roe started to go to it, but he had only taken a few steps when it suddenly sprang up and came at him open mouthed. Again he brought his trusty rifle to his shoulder, and again the bear fell, but only to rise and keep coming. Thus did this desperate battle continue until, with a thrill of horror, the brave hunter heard the peculiar click which announced that he had reached his last cartridge. The ferocious animal, covered with blood and froth, was within fifteen feet of him, and his life hung on his last shot. With a hurried aim, he fired into its wide open mouth. The ball, passing into its throat, broke its neck, and at last it fell to rise no more. Some idea of the danger incurred in hunting this kind of game may be gathered from the fact that every one of those twelve balls struck this bear with sufficient force to knock him down. Several passed through his lungs close to his heart, and many of his bones were broken yet, until the last shot, he possessed life and strength enough to have torn half a dozen men to pieces. He was an unusually large bear, even for Montana, weighing over 1,100 pounds, and was so fat that near ten gallons of oil was rendered from a portion of his carcass. In conclusion it may be said that while Roe still loves to hunt, yet his desire for bear is entirely satisfied.

After crossing the main range of the Rocky Mountains, the road passes for forty miles through a region that is quite sparsely peopled. The face of the country is that of a hilly prairie, covered with excellent pasturage for the herds of cattle which roam over the adjacent hills. The few ranchmen who cultivate the bottom lands of the Little Blackfoot find the soil quite productive.

Frenchwoman's.—The first station is at Frenchwoman's Creek. The creek derives its name from the tragic fate which met a Frenchwoman who kept the stage station here many years ago. One morning she was found murdered, and some hundreds of dollars, which she had hoarded, were missing. Suspicion naturally fell upon the woman's husband, who disappeared at the time of the murder, but he was not captured and brought to justice. The grave of the victim, enclosed by a wooden paling, is seen upon a grassy height, just above the house where the crime was committed, and serves as a pathetic reminder of the event.

Leaving Frenchwoman's, the route follows the winding valley of the Little Blackfoot by an easy down grade to the confluence of this stream with the Deer Lodge River. Here the new town of *Garrison* has been established. It is named in honor of William Lloyd Garrison, the great leader of the anti-slavery cause in the days before the rebellion.

At this point the Utah and Northern Railroad, (narrow gauge,) coming from the South through the Deer Lodge Valley, taps the Northern Pacific Railroad. The Northern Pacific Railroad will place a third rail from its main line, at the mouth of the Little Blackfoot, along the track of the Utah and Northern Railroad to Butte, Montana, a distance of fifty-five miles. This will give the Northern Pacific Railroad access to the most important mining centre of Montana. A third rail will also be laid along the track of the Northern Pacific main line, from Little Blackfoot into

Helena, to enable the narrow gauge cars of the Utah and Northern Railroad to reach the capital of the Territory.

Deer Lodge—(11 miles from *Garrison*; population, 1,200)—derived its name from the abundance of deer that roamed over the broad, open prairie, and from a mound which, on a winter's morning, bore a resemblance to an Indian lodge when the steam issued from the hot spring on its summit. Deer Lodge is the seat of Deer Lodge County, and appears quite attractive, nestled midway in the valley, 4,546 feet above the sea. The town is well laid out, and, with its public square, large public buildings, court-house, jail, churches and educational establishments, makes a good impression. There are three hotels, several wholesale and retail business houses, shops and a weekly newspaper. The town is a general supply and distributing point for several fertile valleys and the surrounding mining districts. Deer Lodge Valley extends fifty miles southward, and is composed of farming and grazing lands. The latter rest on the foot-hills and mountains, while the former are lower down, adjacent to the mouths of the streams. There are remarkable boiling springs in the valley. Many bright mountain trout streams course through its broad expanse, some having their sources eastward in the Rocky Mountain Divide, and others coming from the west through the low, rolling, open country between the Deer Lodge and Bitter Root valleys. Deer Lodge County is noted for the number, extent and richness of its placer mines, and for years it has led the production in placer gold. Among the surrounding mountains, Powell's Peak, twenty miles west of Deer Lodge City, and 10,000 feet in height, is prominent. There are many small lakes in the mountains, which are full of trout, and large game also abounds.

Butte—(44 miles from Deer Lodge; population, 8,000)—is situated near the head of Deer Lodge Valley, and about fifteen miles west of the Pipestone Pass of the main range of the Rocky Mountains, on ground sloping to the south. It is the county-seat of Silver Bow County, and is famous for its quartz mines, which are so largely developed as to make Butte the most important mining centre in Montana. In 1875 the first mill was constructed for working the silver ores of the camp, and the population did not exceed 200. To-day Butte City counts its inhabitants by thousands. Up to within a short time little foreign capital was invested in the mills and smelters of the camp, but it is now beginning to come in. The rapid growth of population and wealth in Butte has few parallels in the mining annals of the country, and the prospect is that within a few years the town will be the most productive mining centre in the United States. A peculiarity of the Butte mines is that, almost without exception, wherever a shaft has been sunk it has paid a handsome profit over and above the cost of working. To the north of the town the ground rises 500 feet higher to the Moulton, Alice and Lexington mines. Besides these mines, or lodes, there are many others, among which are the Shonbar, Bell, Parrott, Gagnon and Original. The veins are true fissures, yielding largely of copper and silver, and assaying well. It is estimated that there are over 300 miles of veins in the district, varying in width from thirty to fifty feet, and developed to a depth of 600 feet. A new smelter is in course of construction by one of the mining companies, which will have a capacity of smelting 500 tons daily. The city is substantially built with large business blocks and fine residences, which, together with its churches, school buildings and hotels, present an attractive appearance. Butte also has two well conducted daily newspapers.

There are many other gold, silver and copper lodes in Silver Bow County which are more or less developed. At present the district embraces the principal mining region of Montana.

From Garrison Westward.—After leaving Garrison, there are fine views of mountain scenery, especially on the left hand, where the snow-mantled peaks of Mount Powell appear. The railroad passes along near the Deer Lodge River, which skirts the heights to the right. The entire region is noted for the richness and extent of its placer mines. Some distance southward are the Gold Hill mountains, where the diggings are very profitable, and the valleys of Rock, Willow and Squaw creeks, streams which flow into the Deer Lodge, have also produced large quantities of fine gold.

The old town of *Pioneer*, once a famous mining camp, is only a short distance south of the track, situated amid a desert waste of hundreds of acres of great boulders and heaps of cobble stones, every speck of soil having been washed away in the search for gold. Pioneer looks ashamed of itself in the midst of the desolation, and has shrunken into a skeleton of a town as compared with its robust appearance a few years ago. Its people hope to increase their trade with the populous and rich quartz-mining camps at Boulder, twelve miles distant southward, or else their town is likely to vanish out of existence.

Onward a few miles, and Deer Lodge River changes its name to Hell Gate River. The valley here rather abruptly narrows, its breadth for seven or eight miles scarcely exceeding a single mile, with mountains on the right hand and bold bluffs on the left, but it again becomes broader where the waters of Flint Creek flow from the south and swell the volume of the river.

At this point a station, called Edwardsville, has been established. A short distance south, on the banks of Flint Creek, there is a farming and trading settlement, known as New Chicago. Still further southward, among the Flint Creek Hills, a distance of twenty miles, is the flourishing mining camp of Philipsburg, with over four hundred inhabitants.

The next station reached is at Bear's Mouth, near Birmingham's Ranch, which was an important stage station before the advent of the railroad. Here passengers between Deer Lodge and Missoula remained over night, and the primitive accommodations and good cooking at Birmingham's were famed in all the region round. The valley now becomes wider, and the surrounding mountains, whose slopes are covered with yellow pine, loom grandly skyward. The bottom lands yield heavy crops of hay, and the gardens of the ranchmen produce potatoes, cabbages and vegetables of enormous size.

Ten miles from Birmingham's the Hell Gate Cañon is entered. This is, however, no narrow mountain pass, as its name would indicate, but, rather, a valley from two to three miles in width, extending a distance of forty miles to the junction of the Hell Gate River with the Big Blackfoot, after which it widens to unite with the valley of the Bitter Root, whereon Missoula stands. The scenery along the Hell Gate Cañon is very fine, often grand. Rock-ribbed mountains rise on either hand, their slopes black with noble specimens of yellow pine, and flecked in autumn with the bright gold of giant tamaracks. The stream itself is deep and swift, quite clear also, except where it receives the murky waters of its many tributaries, which latter in summer are always coffee-colored from the labors of the gold-washers in the mountains. Many islands, covered with cottonwood and other deciduous growths, lie in the crooked channel, adding to the general picturesqueness. Two-thirds of the way down the cañon, Stony Creek, a fine, bold mountain stream, enters from the southwest, after flowing eighty miles through the range between the Deer Lodge and Bitter Root valleys. The water teems with trout. The Big Blackfoot, Hell Gate's largest tributary, comes in from the east, with a valley eighty miles long and varying from half a mile to twelve miles in width, considered one of the finest grazing and agricultural sections in Montana. Many good quartz and galena leads have been discovered in the mountains, and the Wallace District, near Baker Station, is especially promising. At intervals along the cañon, railroad stations have been established, and the immediate occupation of the settlers will be to turn the mammoth pine trees which darken the narrow valley into lumber, preparatory to cultivating the soil.

Beaver Hill, Hell Gate Cañon, near Missoula, Montana.

Beaver Hill—A Legend.—In traveling between Deer Lodge and Missoula, twenty-eight miles from the latter place, at Kramer's Ranch, a remarkable ridge or tongue of land is seen stretching across the valley of the Hell Gate River from the east side, almost in the form of a beaver *couchant*. It is known as Beaver Hill, and it projects so near to the mountains on the west side of the valley as to nearly dam up the river, which is here compressed into a narrow, rocky channel. There is a legend connected with this hill, which is about as follows:

A great many years ago, before the country was inhabited by men, the valleys along the whole length of the river and its branches were occupied by vast numbers of beavers. There was a great king of all the beavers, named Skookum (which in Indian means "Good"), who lived in a splendid winter palace up at the Big Warm Spring Mound, whereon the Territorial Insane Asylum is now situated. One day the king received word that his subjects down the river had refused to obey his authority, and were going to set up an independent government. In great haste he collected a large army of beavers, detachments joining him from every tributary on the way down. On arriving at the great plain now crossed by Beaver Hill he halted his army and demanded of the rebels that they pay their accustomed tribute and renew their allegiance. This they insultingly refused to do, saying they owned the river below to the sea, that it was larger and longer than that above, and as they were more numerous they would pay tribute to no one. The old king was able and wily, and immediately sent for every beaver under his jurisdiction. When all had arrived he held a council of war, and said that as he owned the sources of the great river he would dam it at that point, and turn the channel across to the Missouri. This would bring the rebels to terms below, because they could not live without water. He so disposed of his army that in one night they scooped out the great gulch that now comes in on the north side of Beaver Hill, and with the earth taken out the hill was formed in a night, and so completely dammed up the river that not a drop of water could get through. When the rebellious beavers below saw the water run by and the river bed dry up, they hastened to make peace, paid their tribute (internal revenue tax perhaps), and renewed their former allegiance. So King Skookum had the west end of the dam removed, and ever since that time the river has run "unvexed to the sea." To commemorate the event he had the earth piled up on top of the hill to resemble a beaver in form, and it can be seen either up or down the river a long way. The Indians who first settled up the valley got this legend from the beavers, their cousins, more than a thousand years ago; for in those ancient times they could converse together, and did hold communication until some young and treacherous Indians made war on the beavers for their furs, when the beavers solemnly resolved never to converse with them again, and have steadfastly kept their word.

Missoula—(124 miles from Helena; population, 1,000)—is the county-seat of Missoula County. It is beautifully situated at the western gateway of the Rocky Mountains, on a broad plateau on the north side of the Missoula River, near its junction with the Bitter Root and the Hell Gate, and commands a lovely view of the valley and the surrounding mountain ranges that stretch away as far as the eye can see. This town used to be as isolated and remote a frontier post as could be found in the Northwest, but the railroad has converted it into a stirring, ambitious place, which counts on soon increasing its population from hundreds to thousands. It contains churches, a national bank, a school, a well conducted newspaper and many attractive and substantial business blocks and residences. There are also a large flouring mill, saw mills and a canning establishment, where various kinds of fruit are put up for winter use. Missoula will be an important point on the railroad. The fertile lands of the plain near by, and the large and rich valley of the Bitter Root, already well settled, over eighty miles long, with an average width of about seven miles, besides other agricultural districts to the northward, all make a lively trade. The altitude of this region is about 3,000 feet. The climate is not as cold as in a similar latitude east of the Rocky Mountains, and the soil produces readily a great variety of cereals, fruits and vegetables.

The country surrounding Missoula has been the scene of many fierce conflicts between the Indians. Before the whites inhabited the Territory the Blackfeet Indians ambushed Chief Coriacan of the Flatheads in a defile, fourteen miles north of the city, with a portion of his tribe, and massacred nearly every man. A few years later the Flatheads avenged their chief's death by killing a like number of Blackfeet in the same defile which now bears Coriacan's name.

Missoula County embraces the large and fertile valleys of the Bitter Root and Jocko. Its assessed valuation in 1882 was $2,000,000. Its area of surveyed lands at that date was 600,000 acres, of which one-fourth was under cultivation. The county is heavily timbered, and is rich in mineral and grazing lands. It contains also many beautiful lakes, well stocked with fish and frequented by water fowl. Good trout fishing, as well as

Marent Gulch Bridge.

various other kinds, is obtained in the Missoula, the Bitter Root, Jocko, Lo-Lo, Flathead, Big Blackfoot and Pend d'Oreille rivers, and in Stony and Ashley creeks. The mountain goat is in abundance and can be found in the vicinity.

Fort Missoula, a garrison of the U. S. troops, is pleasantly situated about half an hour's drive from the town in the Bitter Root Valley.

Leaving Missoula, the railroad passes westward across the northern edge of the plain, over a low and well timbered divide, which separates the waters of the Missoula River (the continuation of the Hell Gate) from those which drain into the Flathead. Fourteen miles from Missoula the road enters the Coriacan Defile, and crosses the Marent Gulch by means of a trestle bridge 866 feet in length and 226 feet in height, the construction of which required 1,000,000 feet of lumber. The track follows no valley, but proceeds along the faces of hills, which are covered with fir, pine and tamarack, down into the valley of the Jocko River, where the agency of the Flathead Indians is established.

The Flathead Indian Reservation.—This reservation extends along the Jocko and Pend d'Oreille rivers a distance of sixty miles. It contains about 1,500,000 acres, which, if divided among the 1,200 Flathead, Pend d'Oreille and half-breed Indians who hold the tract, would give 5,000 acres to each family of four persons. A large part of the reservation consists of a mountainous area, with a growth of valuable timber, but there is also a fair quantity of fine grazing land, as well as many well sheltered arable valleys. Mr. E. V. Smalley visited the reservation in the summer of 1882, and gave the result of his observations in the *Century Magazine* for October of that year as follows:

"The Flathead agency is under the control of the Catholic Church, which supports a Jesuit mission upon it, and has converted all of the inhabitants to at least a nominal adhesion to its faith. At the mission are excellent schools for girls and boys, a church, a convent and a printing office, which has turned out, among other works, a very creditable dictionary of the Kalispel or Flathead language. The agent, Major Ronan, has been in office over five years, and, with the aid of the Jesuit fathers, has been remarkably successful in educating the Indians up to the point of living in log houses, fencing fields, cultivating little patches of grain and potatoes, and keeping cattle and horses. The Government supplies plows and wagons, and runs a saw mill, grist mill, blacksmith shop and threshing machine for their free use. There is no regular issue of food or clothing, but the old and the sick receive blankets, sugar and flour. Probably nine-tenths of these Indians are self sustaining. Some persist in leading a vagabond life, wandering about the country; but these manage to pick up a living by hunting, fishing and digging roots, and sell ponies enough to buy blankets, tobacco and powder. But even the best civilized, who own comfortable little houses, with plank floors and porcelain door knobs, got from the Government, like to keep their canvas lodges pitched, and prefer to sleep in them in summer time. Farming is limited to a few acres for each family, but herding is carried on rather extensively. Thousands of sleek cattle and fine horses feed upon the bunch pastures along the Jocko and the Pend d'Oreille, on the Big Camas Prairie and by the shores of Flathead Lake.

* * * Probably there is no better example of a tribe being brought out of savagery in one generation than is afforded by the Flatheads, and their cousins, the Pend d'Oreilles. Much of the credit for this achievement is, no doubt, due the Jesuit fathers, who, like all the Catholic religious orders, show a faculty for gaining an ascendency over the minds of savages, partly by winning their confidence by devoting themselves to their interests, and partly, it may be, by offering them a religion that appeals strongly to the senses and superstitions. These Indians boast that their tribe never killed a white man. They are an inoffensive, child-like people, and are easily kept in order by the agent, aided by a few native policemen. Life and property are as secure among them as in most civilized communities. With them the agency system amounts only to a paternal supervision, providing implements and machinery for husbandry, and giving aid only when urgently needed. It does not, as upon many reservations, undertake the support of the tribe by issuing rations and clothing. Instead of surrounding the agency with a horde of lazy beggars, it distributes the Indians over the reservation and encourages them to labor. It ought to result in citizenship and separate ownership of the land for the Indians. Many of them would now like deeds to the farms they occupy, but they cannot get them without legislation from Congress changing the present Indian policy. Practically, they control their farms and herds as individual property; but they have no sense of secure ownership and no legal rights as against their agent or the chief. Some of them complain of the tyranny of the native police, and of the practice of cruelly whipping women when accused by their husbands of a breach of marriage vows—a practice established, it is charged, by the Jesuits; but in the main they seem to be contented and fairly prosperous. Among them are many half-breeds, who trace their ancestry on one side to Hudson's Bay Company servants or French Canadians—fine looking men and handsome women these, as a rule. They are proud of the white blood in their veins, and appear to be respected in the tribe on account of it; or perhaps it is their superior intelligence which gains for them the influence they evidently enjoy. Shiftless white men, drifting about the country, frequently attempt to settle in the reservation and get a footing there by marrying squaws; but they are not allowed to remain. The Indians do not object to their company so much as the agent."

Flathead Lake, a magnificent sheet of water, twenty-five miles long and six miles wide, is situated on the reservation. The fertile lands about the lake shores are occupied by Indians, whose farming operations are well conducted. The lake is noted for its picturesque scenery. Wooded islands dot its waters, and large, land-locked salmon live in its crystal depths.

Alice Falls, Montana.

The Flathead or Pend d'Oreille River issues from the lake, and flows with strong current, and with a fall of fifteen feet at one point in its course, to its junction with the Jocko, thirty miles below. Near the Mission of St. Ignatius are the cascades, known as the "Two Sisters," a visit to which Gen. Thomas Francis Meagher, Acting Governor of Montana, who was drowned at Fort Benton in 1867, eloquently described. He wrote:

'Topping a low range of naked hills, we had a sight which made the plastic heart of the writer dilate and beat and bound and burn with rapture. Beyond there, walling up the horizon, were the Rocky Mountains, rearing themselves abruptly from the plains and valleys—no foot-hills, no great stretches of forest, to detract from the magnificent stature with which they arose and displayed themselves unequivocally, with their bold and broken crests, with their deep and black recesses, with their borders of white cloud in all their massiveness and stern, co'd majesty, in the purple light of a mid-summer evening, the calmness and the glory of which were in full consonance with the dumb, gigantic features of the scene. Right opposite, leaping and thundering down the wall of a vast amphitheatre that had been scooped out of the mountains, was a torrent, bounding into the chasm from a height of fully two thousand feet, but looking as though it were a bank of snow lodged in some deep groove, so utterly void of life and voice did it appear in the mute distance. A mass of trees blocked the bottom of the amphitheatre; and following the torrent which escaped from it after that leap of two thousand feet, thousands and tens of thousands of trees seamed the valley with a dark green belt, all over which the hot sun played in infinite reflections and a haze of splendor. The path to this chasm lies through a dense wood, the beautiful and slender trees in which are closely knitted together with shrubs and briers and snake-like vines; while vast quantities of dead timber and immense rocks, slippery with moss, and trickling streams, thin and bright as silver threads, encumber the ground and render it difficult and sore to travel. There are few tracks there of wild animals, and all traces of the human foot are blotted out, so rarely is that solitude visited even by the Indian.

'As we neared the foot of the Elizabeth Cascade—for such was the name given to the headlong torrent—great was our surprise to find another torrent equally precipitous, but still more beautifully fashioned, bounding from the edge of the opposite wall, and as a jutting rock, sceptered with two green trees of exquisite shape and foliage, dispersed its volume, the torrent spread itself into a broad sheet of delicate foam and spray, white and soft, and as full of light and lustre as the finest lace-work the harvest moon could weave upon calm waters. This cascade is completely hid from view until one stands close under it, and the Fathers of the Mission, strange to say, knew nothing of it until our explorers told them exultingly of their discovery. To this they gave the name of the Alice Cascade, christening them both the Two Sisters."

The railroad follows the beautiful valley of the Jocko River to its confluence with the Flathead, forty-four miles from Missoula. The Flathead for the next twenty-five miles, until its waters are united with those of the Missouri, is now called the Pend d'Oreille River. Keeping along the left or southern bank of this stream for seventeen miles, the road sweeps around a grand curve and crosses to the right bank over a fine truss bridge, which, with its approaches, is about 800 feet long. Eight miles beyond the crossing, the muddy waters of the Missoula, pouring in from the south, mix with the bright flood of the Pend d'Oreille, and the united streams now take the designation Clark's Fork of the Columbia. This name is retained, except where the river widens out into Lake Pend d'Oreille, 100 miles westward, until the waters mingle with those of the Columbia River, in the British Possessions northward.

Paradise Valley and Horse Plains.—Two small and charming valleys soon appear to vary the fine mountain views. They are Paradise Valley and Horse Plains, both celebrated among the Indians as wintering places for their ponies. Paradise Valley is seven miles westward of the junction of the rivers. It is two by four miles in extent, and well deserves its name. Six miles beyond is Horse Plains, a circular prairie, six miles across, containing a township of fertile land, situated in the midst of very wild scenery. High mountains stand around and lend the warmth of spring, while their own sides are white with snows. These valleys are the only spots on the immediate line of the railroad for over a hundred and fifty miles that invite cultivation. The oldest inhabitant of this region is one Neptune Lynch. He drifted hither almost twenty years ago, and was content to own a few cows and let them roam the wilderness. The small herd of sixty cows grew and throve. They summered in the mountains and wintered in the valleys, where snow seldom falls over four inches in depth. Lynch's stock, which roam for a hundred miles, have made him and his sons rich. The land of Horse Plains produces everything desirable in a northern latitude, under irrigation, but in some seasons irrigation is not needed.

Leaving Horse Plains and Crossing Clark's Creek, with Lynch's Buttes visible to the right, the railroad continues westward along the right bank of the river through an unbroken mountain region, which affords magnificent views at every turn. The mountains tower on either side. There is no bench land, much less any fertile bottoms, though sometimes level spots of a few acres are heavily timbered. Room is not always found for the track, which is often blasted out from the points of the hills. The grand surroundings of the route at times produce remarkable effects,

Weeksville and its Vigilantes. While the railroad was under construction, temporary towns sprang up along its course to disappear as the work moved on. The structures were either of boards or canvas, rudely put together and easily removed. Weeksville, now called *Kitchen's*, one of many such towns, about ten miles west of Horse Plains, was rendered both famous and infamous by the deeds that happened there. These towns were not only places of trade, with stores and eating-houses, but contained drinking dens and gambling hells of the worst sort, where murder was often the ending of affrays. It was dangerous for a person to be out through the region if he was supposed to have money. That many a foul deed took place despite all the efforts of the company to the contrary, is attested by the finding of dead bodies by the wayside or floating down the turbulent stream. The establishing of a vigilance committee was the natural result of all absence of legal safeguard. Several desperadoes were summarily hung, and afterwards the camps were relieved of the presence of the ruffians. Neither a gambler nor a saloon keeper is necessarily a murderer, and the most respectable and peaceable of this class joined the movement to get rid of the thugs. Close by the track where Weeksville was in the winter of 1882–3, there is yet a graphic though rude memento of that rough time in three nameless graves that mark where as many lynched outcasts were buried. The only token that marks the spot is that at one of these rude mounds two crutches are stuck in the earth. One of the victims of justice attempted to commit a robbery a few weeks before his execution, and in the attempt the accidental discharge of his pistol inflicted a wound. He had recovered sufficiently to walk on crutches when the vigilantes took the field. They traced his career and decided to make him an example. He was hung, and the irony of his executioners erected to his memory a suitable memento by sticking his abandoned crutches over his timely grave.

If the history of the construction of the railroad and the attending circumstances could be written, many a tragedy would be recorded. Chinamen by the thousand, and some of the worst character (for there are as many grades of Chinese as of Caucasians), worked on this long stretch of road through what was a wild, uninhabited region. There was no law, no civilization. Chinese are inveterate gamblers, and opium smoking is a vice as terrible as drunkenness. Many a poor wretch who had received his wages was treated by the outcasts of his own race in the same manner as the white ruffians served their victims. The unsuspecting laborer was robbed of his earnings, murdered, and the icy flood of Clark's Fork was always ready to receive and bear away the body of a victim.

Grand Scenery.—There are stations every few miles along Clark's Fork, but no towns, or reason for them, unless some points may in time be connected with the better regions known to exist to the northward towards the British line.

Everywhere along this stretch of turbulent water there is magnificent scenery. Cottonwood grows close to the river, and firs and pines clothe the benches and mountain sides, except where the latter are so nearly vertical that forests cannot grow. Magnificent vistas are presented as the train moves along, changing and wearing new forms at every turn. The mountains are conical and sometimes vertical, as where the river has cut through them with tremendous force. The constant succession of towering hills, grouped in wild array, is never wearying, and is sometimes startling in effect, as when some tributary from the north or south tears its way to the greater stream, and offers a vista, reaching far through the deep-worn cañon or ravine, along which the heights are ranged as far as eye can see. One of the most striking of these side effects is where Thompson's River comes in from the north, and you look up the long and sharp ravine to catch a momentary glimpse, from the trestle bridge, of the foaming waterfall and the heights that wall it in.

Thompson River Station, thirteen miles from Kitchen's, and a mile or so west of the point where the track crosses that stream, is placed on a plateau which is planted by nature with scattering pines, and surrounded by a grand cordon of mountains. For several miles here the scenery is very fine, equal to any views to be seen upon the road.

Reaching the second crossing of the Clark's Fork, there is seen a navigable stretch of water that was utilized by placing a small steamer on it at the time the railroad was under construction. East of the second crossing the mountains close in upon the view, often abruptly. West of it the valley widens. There is no land to style it a valley, but the gorge is wider and the river less turbulent. The scenery has the same features, but in rather quieter lines, as the heights do not crowd the river so much. The road is now on the south side of the stream. West of Second Crossing, about ten miles, the track follows a high bench, and a view is shown of the river where its waters have cut a deep channel far below. Mountains on the north stand imminent, and make a striking picture.

Good Hunting and Fishing.—Another feature of this mountain region, which is likely to attract the attention of lovers of sport, is that abundance of game is found among all the ranges. There is no other region that can surpass it for the presence of wild and game animals, as well as birds and fish. Bears are very common; elk, caribou or moose haunt these mountains, and deer of various kinds abound. There are many of the fur-bearing animals, such as otter, beaver and mink, while grouse, pheasants, ducks, geese and other fowl are plentiful in their season. The waters abound in the finest trout of various varieties, from the little speckled beauties of the mountain rills to the great salmon trout found in the larger streams and lakes.

Light Snow Fall.—Until the winter of 1882-83 it was not even approximately known to what extent the snow fell on the Clark's Fork Division of the railroad. As this rough mountain region was occupied by the constructing force during the entire winter season, ample opportunity was given for observation. It was found that the snow belt reaches from Sand Point, on Lake Pend d'Oreille, to nearly the second crossing of the Clark's Fork, a distance of forty-one miles. The greatest depth of snow upon the ground was seven feet, and the average depth was six feet. This snow was so light that it did not pack, and there was no difficulty in keeping the track clear by close watchfulness and a free use of the snow plow. Beyond the second crossing of Clark's Fork, up to the Jocko Indian Reservation, there was absolutely no snow, and the few Indians along this part of the line say that it rarely falls in sufficient quantity to cover the ground. Early in March, 1883, the grass was quite green, and the Indians had begun to plow.

After coursing along the northern and southern banks of the Clark's Fork of the Columbia for a hundred miles, the views of mountain and forest sometimes broadening, sometimes narrowing, and the river alternately showing a wooded reach of smooth water and a stretch of tumbling breakers, the mountains again crowd together near Cabinet Landing. The stations on the next fifty miles after passing Thompson's Prairie, are *Belknap*, *White Pine*, *Trout Creek*, *Big Cut*, *Summit* and *Noxon*. These stations are either for the convenience of the railroad employees or for the shipment of lumber, and in every other aspect are at present of not sufficient importance to be described. At several points on the line the track is carried across lateral streams by massive trestle bridges, the one over the deep gorge of Beaver

Creek being especially noticeable from its height and graceful curve. These frequent bridges, as well as many deep cuttings through the spurs of the mountains, attest the difficulties which the engineers were required to surmount in constructing the line.

ALONG THE CLARK'S FORK.

Cabinet Gorge, on Clark's Fork.

Pend d'Oreille Division.

FROM HERON TO WALLULA.—DISTANCE, 269 MILES.

Heron—(274 miles from Helena; population, 300.)—This is a new town built upon a plateau in the midst of a dense forest. It is wholly a creation of the railroad. Here the Rocky Mountain Division ends and the Pend d'Oreille Division begins. Heron has a round-house and a repair shop, and its people are nearly all in the service of the road. It is expected that important lumber industries will be developed here in the course of time, and that the place will become the supply point for mining industries to be established in the surrounding region.

Cabinet Landing.—At this point, five miles east of Heron, the river is confined in a rocky gorge, through which it dashes at tremendous speed. The columnar rocks that hem in the torrent are from 100 to 150 feet in height, their brows crowned with pines, and the romantic wildness of the gorge is of surpassing beauty. The bold, fluted pillars of rock are not unlike those of the "Giant's Causeway" in Ireland. Cabinet Landing derives its name, in part at least, from the fact that here the Hudson's Bay Company, in carrying up goods by boat from the foot of Lake Pend d'Oreille to Horse Plains, was compelled to make a portage. From Cabinet Landing the train runs through solid rock cuttings, the walls of which tower far above the rushing, tumbling stream below. *Clark's Fork*, a station on the confines of Idaho, eight miles from Cabinet, is next passed, and ten miles further the pleasant town of *Hope*, on the strand of Lake Pend d'Oreille, is reached.

Idaho.

The Northern Pacific Railroad passes over a very narrow strip of northern Idaho—scarcely a degree of longitude—between the eastern end of Lake Pend d'Oreille, and to a point near Spokane Falls, W. T. Idaho is bounded on the east and northeast by Montana and Wyoming, from which Territories it is separated by the winding chain of the Bitter Root or Cœur d'Alène Mountains. On the south it follows the forty-second parallel along the line of Utah and Nevada. On the west lie Oregon and Washington, and on the north the British Possessions. Idaho is embraced between the forty-second and forty-ninth parallels of latitude, and between the 111th and 117th meridians of longitude, west of Greenwich. Its area is 86,294 square miles, or 55,228,160 acres. The northern part of the Territory is quite mountainous, some of the highest altitudes reaching 10,000 feet. Mountain and valley alike are covered with a dense growth of coniferæ. The principal ranges are the Bitter Root and the Salmon mountains, the latter traversing the central portion of the State. South of this mountain range, stretching nearly across the Territory, is the Snake River Plain, the surface of which is either level or gently undulating. Still further south is an elevated plateau, which merges in the southwest into an alkaline desert. Idaho is on the whole well watered. Its principal stream is the Snake or Lewis Fork of the Columbia, which, with its many affluents, drains about five-sixths of the Territory. This stream, generally confined within high walls of basalt, pursues a tortuous and tumultuous course, from its sources in Wyoming, of about 1,000 miles, interrupted by many falls of considerable height. It is only navigable from a short distance above Lewiston, near which city it leaves the Territory, to its junction with the Columbia River at Ainsworth, less than 100 miles distant. The principal tributaries of the Snake River are the Salmon, the Boisé, the Gwyhee and the Clearwater, the Salmon River draining the central part of the Territory.

The arable lands of Idaho are estimated at ten per cent. of its area. There are fine small valleys in the northern part on all the streams flowing into the Snake River from the east, with an abundance of water. In the south there are also good valleys, which could be cultivated by irrigation. The grazing lands of Idaho cover a great area, especially in the southern part of the Territory. All the level country of the Snake River Plains is valuable for pasturage, as well as the mountain ranges to the south and southeast, which are covered with bunch grass.

The Territory was organized in 1863, having been cut off from Oregon, although a part of it was subsequently given to Montana. The mineral resources of the Territory are very great, but as yet they have been only slightly developed. The principal yield has been from the placers along the Snake and other rivers, which has amounted to about $75,000,000 since operations were begun. There was no railroad in Idaho before 1880, at which date the Utah and Northern narrow gauge railroad ran its line through the southeastern part, and has since extended its system to Montana. The Oregon Short Line, a branch of the Union Pacific Railroad, diverging from Granger, in Wyoming, is now also under construction to meet the Oregon Railway and Navigation Company's system which is in process of extension beyond the Blue Mountains, to Baker City. The gradual failure of placer mining has very much stimulated the prospecting for lodes of gold and silver, and valuable discoveries of these metals have been made in all the mountain ranges, requiring only better transportation facilities for their development.

The assessed valuation for 1882 was $9,108,450. The population is estimated to be near 50,000, exclusive of the Indians, who number about 5,000. These Indians consist of the Nez Percés, Bannocks and Shoshones. The former, numbering 2,807, have a reservation of 1,344,000 acres on the Clearwater, near Lewiston, toward the northern part of the Territory. The two latter tribes, numbering 1,500, jointly occupy a reservation of 18,000 acres in the southeastern part of the Territory, on the Snake and Portneuf rivers. There is also a reservation near Lemhi, in the Salmon River Mountains, where 677 Indians are reported as having their homes.

Pend d'Oreille Division.

[Continued from page 205.]

Lake Pend d'Oreille.—This beautiful lake may be likened to a broad and winding valley among the mountains, filled to the brim with gathered waters. Reaching the lake, the railroad crosses the mouth of Pack River on a trestle one mile and a half in length, and skirts the northern shore for upwards of twenty miles. The shores are mountains, but wherever there is a bit of beach it is covered with dense forest. The view of the lake from the car windows, with its beautiful islands and its arms reaching into the surrounding ranges, is superb. The waters stretch out south, and fill a mountain cove to the southwest before those of the Clark's Fork meet them. From this point the river makes the lake its channel, and passes out at the western end on its flow northward to meet the Columbia, just over the boundary line in British Columbia. The whole length of the lake, following its curves and windings, must be nearly sixty miles. In places it is fifteen miles wide and in others narrows to three miles. It is probable that the steamer now on the lake will ply for the pleasure and relief of travelers. The railroad goes north of the lake, by a circuitous route. It is possible to put on stages to connect from the southwest at some point above Rathdrum, a town on the line of the road distant only six miles from the west shore. Here the steamer could take passengers across the lake, and for some distance up Clark's Fork, making a saving of twenty-five miles before again connecting with the train, thus affording a relief from the monotony of railroad travel, and at the same time a fine opportunity to view the scenery of the lake.

The circuit of the lake shore is full of surprises. The mountains are grouped with fine effect, and never become monotonous. Along the lake the most permanent features of civilization are the saw mills, which supplied material for railroad construction, and are now employed manufacturing lumber for shipment. The forest is interminable, but where the mountains are abrupt the trees do not grow large enough and clear enough

Lake Pend d' Oreille Idaho.

to make good lumber. The benches and levels along the streams are generally thickly studded with giant pines or firs, and these trees also tower in the ravines. These spots of good timber are selected as sites for saw mills, and the carrying of lumber will be an important branch of traffic. The Northern Pacific road reaches its farthest northern limit at Pend d'Oreille, and thence turns south and west.

Kootenai—(28 miles from Heron.)—At this station the Pack River enters Lake Pend d'Oreille, and from here an old fur-trading and mining trail leads to the Kootenai River, a distance of about thirty miles. The Kootenai is an eccentric stream, running first south and making a long bend, and afterwards flowing due north far into British territory. The Kootenai is navigable for 150 miles, for 100 miles of which it expands into a deep, narrow lake. A company of Portland business men has recently placed a steamer upon this river, and design to open a regular route of travel from the Northern Pacific Railroad northward, by way of the Kootenai, to the Canada Pacific Railway, the purposes being chiefly to develop the mining and agricultural resources in the valley of the Kootenai.

Sand Point (38 miles from Heron)—on the shore of Lake Pend d'Oreille, was a place of importance during the time of construction, and probably will retain its advantage to some extent, as it connects with the country on the north. It is also a good point to lie over for a day's hunting, or for catching some of the trout with which the lake abounds. *Algoma* station is six miles beyond.

Cocolala—(51 miles from Heron.)—This station derives its euphonious Indian name from the bright sheet of water which lies near the track. The lake is several miles long, but not wide. On approaching it a charming view of wave, wood and mountain will be caught. But we are passing out of Wonder Land. Mountains no longer seem to overtop us. The train sweeps on towards the southwest, following a natural pass between the ranges, presently entering a valley, a few miles wide. There is no settlement along the road, and no cultivation. The forests sometimes break away and give space for open country, but there is little except continuous woods. The only improvements we see are the railroad stations every few miles, and occasionally a saw mill. These are the pioneers of civilization in the Northwest. The stations have musical Indian names, such as *Chilco* and *Calispel*, and little else. Continuing southwestwardly, the road clings to a mountain side, and reaches *Lake Prescott*, which is a deep, dry hole in summer, and filled with water in winter and spring. Nineteen miles beyond, the Spokane Valley is entered near

Rathdrum—(80 miles from Heron.)—This is the first place of importance reached after leaving Lake Pend d'Oreille. It lies at the northern edge of the Spokane Plain, the richest lands of which are near the town. Rathdrum is a growing place, and is likely to develop into a thrifty town. It is the nearest point on the railroad to the military post of Fort Cœur d'Alène, ten miles distant, on the shores of the lake of the same name. A good hotel for tourists and summer residents has been erected at Lake Cœur d'Alène, immediately adjoining the beautiful park-like ground of the military post. The lake affords excellent opportunities for fishing and boating, and the climate is peculiarly clear and healthful in the summer months.

Nine miles beyond Rathdrum the Idaho line is crossed, and we are in Washington Territory.

Lake Cœur d' Alène, Idaho.

Washington and Oregon.

The prominent features of Washington Territory and of Oregon are so much alike, that a descriptive outline of the topography, soil, climate and resources of the entire region may well be grouped under one heading.

Washington Territory lies between the parallels of 45° 32' and 49° N., and the meridians of 117° and 124° 8' W. Its boundaries are—north by British Columbia, east by Idaho, south by Oregon and west by the Pacific Ocean. The Territory ranges from 200 to 250 miles in length, and its greatest breadth from east to west is about 360 miles. Its area is 69,994 square miles, or 44,796,160 acres. Washington Territory was organized in 1853, and at that time included much of what is now Idaho. The population is estimated at 110,000.

Oregon lies between the parallels of 42° and 46° 18' N., and between the meridians of 116° 33' and 124° 25' W. On the north it is bounded by Washington Territory, on the east by Idaho Territory, on the south by Nevada and California and on the west by the Pacific Ocean. The average width of Oregon, from east to west, is 350 miles, and from north to south 275 miles. Its area is 95,274 square miles, or 60,977,360 acres. The population of the State is about 225,000.

The Cascade Mountains, a broad volcanic plateau, with many lofty, snow-clad peaks, rising high above the general level, divide both Washington Territory and Oregon into two unequal parts, which differ widely in surface, climate and vegetation. Westward of this mountain chain, from forty to seventy miles, is still another and lower range, lying along the ocean shore, known as the Coast Mountains.

Between these two mountain ranges spreads out a great basin, about 400 miles in length, which is broken up into many well watered valleys all of which are fertile, and some of great size, the largest being the Willamette, Umpqua and Rogue River valleys, in Oregon, and the valuable timber area of Puget Sound, in Washington Territory. The entire region west of the Cascade Mountains, including the slopes of these elevations, is covered with dense forests, mainly of coniferæ, which constitute a large source of wealth, especially in the Puget Sound district.

The climate of this section is mild and equable, with slight ranges of temperature, showing a mean deviation of only 28° during the year, the summer averaging 70°, and the winter 38°. There is an abundant rain-fall and the wet and dry seasons are well marked. The rains are more copious in December, January and March than at any other time. But the rain falls in showers rather than continuously, with many intervals of bright agreeable weather, which often last for days together. Snow rarely falls in great quantities, and it soon disappears under the influence of the humid atmosphere. During the dry season the weather is delightful. There are showers from time to time, but the face of the country is kept fresh and verdant by the dews at night and occasional fogs in the morning. The soil of the valleys of western Washington and Oregon is generally dark loam, with clay subsoil, and in the bottom lands near the water courses are rich deposits of alluvium. These soils are of wonderful productive capacity, yielding large crops of hay, hops, grain, fruits and vegetables.

The area east of the Cascade Mountains, by far the larger portion of Washington and Oregon, presents features in marked contrast to those which have been already outlined. This is not only true of climate, but also of soil and topography, fully warranting the popular division of the country into two sections, known as the coast region and the inland region, which are essentially dissimilar in aspect.

The area east of the Cascade Mountains extends to the bases of the Blue and Bitter Root ranges. A narrow strip on the north is mountainous and covered with forest; but the greater portion embraces the immense plains and undulating prairies, 150 miles wide and nearly 500 miles long, which constitute the great basin of the Columbia River. Within the limits of this basin are a score of valleys, many a one of which is larger than some European principalities, all of which are well watered and clothed with nutritious grass.

In the eastern section the temperature is decidedly higher in summer and lower in winter than in the western section—the average indicating respectively 85° and 30°. The rain-fall is only half as heavy, but it has proved sufficient for cereal crops. From June to September there is no rain, the weather being perfect for harvesting. The heat is great, but not nearly so oppressive as a much lower grade would be in the Eastern States, and the nights are invariably cool. The winters are short, but occasionally severe. Snow seldom falls before Christmas, and sometimes lies from four to six weeks, but usually disappears in a few days. The so-called "Chinook," a warm wind, which blows periodically through the mountain passes, is of great benefit to the country. It comes from the southwest across the great thermal stream, known as the Japan current, and the warm, moist atmosphere melts the deepest snow in the course of a few hours.

The soil is a dark loam, of great depth, composed of alluvial deposits and decomposed lava overlying a clay subsoil. The constituents of this soil adapt the land peculiarly to the production of wheat.

Agriculture is the leading industry at present, and wheat is the principal product of the entire country. Its superior quality and great weight have made it famous in the grain markets of the world, and insures for it the highest price, the quantity exported in 1882 being about 10,000,000 bushels. Oats and barley also yield heavily. Hops are a very important product, and widely cultivated in the Willamette Valley, Oregon, and in the Puyallup and White River valleys, on Puget Sound, and in Yakima County, east of the Cascade Mountains, W. T. Vegetables of every variety, and of the finest quality, are produced. Fruits of many descriptions, all of delicious aroma and flavor, grow to a remarkable size. Among them are apples, pears, apricots, quinces, plums, prunes, peaches, cherries and grapes. Strawberries, raspberries, blackberries, gooseberries and currants are also abundant.

An important industry is the raising of cattle, sheep and horses. This is only second to agriculture, and is pursued in all parts of the Pacific Northwest. The horses are of excellent race and excel in speed. Sheep husbandry has proved very profitable, especially among the Blue Mountain ranges, and the aggregate wool clip for export in 1882 amounted to eight or ten million pounds.

It would scarcely be possible to exaggerate the extent and value of the forests. East and west of the Cascade Mountains there are large tracts of timber lands. The Blue Mountains and eastern slopes of the Cascades are thickly clothed with pine timber, and west of the Cascade Mountains there is an inexhaustible supply. Perhaps the finest body of timber in the world is embraced in the Puget Sound district. The principal growths are fir, pine spruce, cedar, larch and hemlock, although white oak, maple, cottonwood, ash, alder and other varieties are found in considerable quantities. The shipment of the product of the saw mills of Puget Sound alone, in 1881, amounted in value to nearly $2,000,000.

The mineral wealth of Oregon and Washington Territory is large and diversified, but the mining industry has not yet been fully developed. Gold was discovered as early as 1851 in southern Oregon. Some time afterward auriferous gravel was found in large quantities in eastern Oregon, and at various times placer and quartz mining has been carried on in the extreme southern part of the Cascade Mountains in Oregon. Washington Territory likewise has yielded more or less of the precious metal. Coal takes a foremost rank among the mineral resources of the country. Immense beds of semi-bituminous and lignite coal are found west of the Cascade Mountains. This mineral exists in Oregon in different localities, but the coal fields of Washington Territory are far more extensive. The principal mines are at Newcastle, near Seattle, and also on the Puyallup and Carbon rivers. Iron ores—bog, hematite and magnetic—exist in great masses, and are found in both Oregon and Washington Territory.

All the rivers of Oregon and Washington flow into the Pacific Ocean, the largest of which, the Columbia, is navigable for a distance of 725 miles. The Willamette River is next in size, and may be navigated by the largest ocean steamships and sailing vessels as far as Portland, 112 miles from the mouth of the Columbia River, and by steamers a distance of 138 miles beyond. The Snake River comes next in importance, and there are many other streams navigable for long distances.

Puget Sound is a beautiful archipelago, covering an area of over 2,000 square miles. Its waters are everywhere deep and free from shoals, its anchorage secure, and it offers every facility that a great commerce will demand.

There are several commodious harbors for vessels of light draft on the coast line, exclusive of those found at the mouths of the several rivers. At these places a thriving trade is carried on in lumbering, coal mining, fishing, oystering, dairying and agricultural products.

These waters abound in fish, of which many varieties are of great commercial value. Particularly is this the fact with regard to salmon. Extensive establishments for canning are carried on at several places. Especially is this the case on the Columbia River, where the business of salmon packing is one of the principal industries. The far-famed reputation which

Spokane Falls, Washington Territory

the Columbia River fish has acquired secures it a large market in the Eastern States, and it is sold extensively in Australia, England and other European countries. The value of this product shipped from the Columbia River in 1882 amounted to $2,729,500.

The varied and valuable natural resources of Washington and Oregon are now attracting large numbers of immigrants.

Pend d'Oreille Division.

[Continued from page 210.]

The Spokane Valley and Lake Cœur d'Alene.—One of the most singular districts of this country is the Spokane Valley. It is thirty miles long and three to six miles in width, surrounded by the western ranges of the lower Cœur d'Alène or Bitter Root Mountains. The Spokane River rises in Cœur d'Alène Lake, close under the timbered mountains, in Idaho, about ten miles south of the railroad. The lake extends south at least forty miles, and has long arms reaching in among the mountains. A rich agricultural region lies close to it on the west, in great part contained in the Cœur d'Alène Indian reservation. The Indians have always been at peace. Under the religious control of the Roman Catholic Church, they have been well taught, and have become civilized, so that they are self supporting. They marketed 30,000 bushels of wheat in 1881, and have extensive timothy meadows on the bottoms of the St. Joseph and Cœur d'Alène rivers, which drain the mountains for 100 miles into Cœur d'Alène Lake.

A proof of the degree of civilization which has been reached by these Indians, quite as strong as the fact that they have a surplus of agricultural products to sell for cash, is that Old Sulteas, their chief, has a pair of well matched horses to his comfortable carriage, and loans money at Spokane Falls on good security for two per cent. a month. These Indians have an earthly paradise for their home. They rent the meadows for fifty miles along the rivers named, and have a monopoly of the forests, so make the settlers adjoining the reservation pay them tribute.

The rivers that drain the western water-shed of the Cœur d'Alène Mountains pour immense volumes into the lake, but the Spokane River, the lake's only outlet, is but a brook in size, with no tributaries of importance. Still, thirty miles below the lake, this stream becomes a roaring cataract at the town of Spokane Falls. The theory is advanced that the region around the lake and all the upper Spokane Valley consists of a deep gravel deposit. Time has made for the lake a water-tight bottom, but a well, dug within a rod of its shores, will not furnish water, and no well can be dug in all the Spokane Valley. The water furnished by the mountains soaks through this immense bed of gravel, making Spokane River, in its upper reaches, so puny a stream. Eight miles below the lake there are the Little Falls, where the river flows between rocks very close together. Thirty miles below the gravel deposit ends, and basaltic shores close in upon the stream. Gradually, as the lower valley is reached, the river is increased in volume as the flow is forced to the surface, and at the falls it is all gathered well in hand, and makes a tremendous leap with a force far greater than would be believed after seeing the Little Falls.

In spite of the gravelly character of the entire plain, there are many large patches and strips where a rich soil has been deposited. This is especially true of the upper end of the valley, and the people near Rathdrum are raising good vegetables and other crops. The railroad runs diagonally through this strange section, and soon reaches the fertile, well wooded and well watered regions of the Palouse and Upper Columbia, into which there has been a heavy immigration for the past two years.

Spokane Falls—(108 miles from Heron; population, 1,600).—This is the first point of importance reached in Washington Territory. It has, in some remarkable respects, more claims to consideration than any other place east of the Cascades. Its situation—upon the gravelly plains just above where Hangman's Creek joins the Spokane River—is very beautiful, looking out upon the hills, with the grand, roaring water-fall in its midst. Spokane Falls is the oldest town in the northeast of Washington Territory, the only one that preceded the railroad. Enterprising men were early attracted to the place, not alone by its natural beauty, but also by the wonderful water power, so easy of control and so abundant in a country that has very few water privileges. It will never be necessary to pave the streets of this city, nor will its people be troubled with mud. Its thoroughfares are macadamized sufficiently by Nature. There is no doubt that Spokane Falls will become a manufacturing city of great importance. There are already in operation extensive flouring mills, and saw mills that supply lumber for use all along the road. Wheat is hauled here from far and near. Logs are cut in the rivers of the Cœur d'Alêne Mountains, 100 miles away, run down the rivers to the lake, and thence down the Spokane to these falls. Wood working factories are in operation, and the place will become more and more the home of thriving industries as years go by. The early settlers who located here and held on to the town-site in faith have become wealthy. Spokane is the most bustling and active place in all the region round. It has large hotels, handsome stores, good schools and churches, newspapers, railroad buildings, and a spirit of enterprise that will make the most of its opportunities. In addition to its other advantages, Spokane also has a good farming region within reach in all directions, and a trade with Fort Colville, Fort Spokane and other districts to the far north.

The falls, seen when melting snows swell the flow and the banks are brimming with the hurrying flood, are a sight never to be forgotten. Basaltic islands divide the broad river, and the waters rush in swift rapids to meet these obstructions. A public bridge crosses from island to island. The width of the river is nearly half a mile. There are three great streams curving towards each other, and pouring their floods into a common basin. Reunited, the waters foam and toss for a few hundred yards in whirling rapids, and then make another plunge into the cañon beyond. Standing on the rocky ledge below the second water-fall, and looking up the stream, a fine view is obtained of the wonderful display of force. All things are weak and trivial compared with the tremendous torrent that heaves and plunges below, and the grand cascades that foam and toss above. Eternal mist rises from the boiling abyss, and sunshine reveals a bow of promise spanning the chasm.

Medical Lakes.—A few miles to the north of the road, almost equidistant between Spokane and Cheney, there is a group of five lakes from one to three miles long. Three of these lakes, having great depth, are very strongly impregnated with alkaline salts, and their water has remarkable curative properties. One in particular attracts hundreds of invalids, especially persons affected by rheumatism, skin diseases and nervous complaints. Many undoubted cures of a remarkable nature are recorded. This medical lake, *par excellence*, has a medium strength of salts, while another has a very strong impregnation, and the third is very weak. The region is delightful, and can be made a very pleasant resort. The proprietors of the town Medical Lake—are doing what they can to accommodate the public by building hotels and erecting bathing establishments. The country people come and pitch their tents and take their baths as they choose. Enterprising men are evaporating the water in heated pans, and so procure the salts for sale.

The early history of this lake is this: A Frenchman, named Lefevre, who was sorely afflicted with rheumatism, was tending sheep around the shores of the lake. He found that after washing the sheep in the lake water his rheumatism was less painful, so he began to bathe his shrunken limbs, for one arm was wasted away and was carried in a sling. The result was a perfect cure of the rheumatism, and restoration of the wasted arm to its natural size. Lefevre still lives at Medical Lake in perfect health, no longer a poor shepherd, for the increase in value of lands from the discovery of the medical properties of the water has made him independent.

Passing the unimportant station of *Marshall*, eight miles beyond, the train arrives at

Cheney—(124 miles from Heron; population, 1,200.)—This is a thriving town that had only a single log house in October, 1880. The railroad company has erected good buildings at all the prominent points, and Cheney has a handsome station-house, with room for a branch office of the Land Department. Close by the station is the Oakes House, a large hotel, built to supply the growing needs of travel. The town is situated in a scattered pine forest, on a spur of the Cœur d'Aléne Mountains, about 2,300 feet above the sea level, and slopes toward the south. On the height facing the town is a handsome academy, with 200 names upon its roll, erected by the beneficence of Benjamin Cheney, Esq., of Boston, after whom the town is named, who, in appreciation of this compliment, gave ten thousand dollars, to be used in building the school. Around Cheney, in all directions, there is a solid agricultural country. The industries of the town are various. There is a steam flouring mill close to the depot. The business street is well lined with stores. There are banks, two newspapers, several churches, and everything to make up a flourishing town. Cheney is constantly growing, and must always be a good point, as it has the agricultural country to support it. Connections are here made for Fairweather, Cottonwood, Medical Lake and other small towns. Passing the small station at *Stevens*, the next stop is at Sprague, twenty-four miles distant.

Sprague—(148 miles from Heron; population, 1,100.)—This is a place of importance, being the head-quarters of the Pend d'Oreille Division of the railroad. Here the railroad company has a large building for its offices, workshops and a round-house, and employs several hundreds of workmen in car building, repairing, etc. Sprague is also within easy reach of good agricultural country in all directions, and does a flourishing trade. The place almost rivals Jonah's gourd, that came up in a night. It was a vacant space in the autumn of 1881, but now it is a favorite resort of immigrants, being situated centrally to the fertile regions north of Snake River and along the railroad.

A singular fact in relation to all this upper country is, that the railroad, for hundreds of miles, either follows the banks of rivers or the dry beds of old water courses. The traveler does not see any good, arable land as he journeys through it. At Sprague, looking eastward, there is a range of purple hills a few miles distant that are the western boundary of the fertile Palouse country. The level land between these heights and the railroad is rocky, with frequent ponds, and *Lake Colville*, two miles west of Sprague, lies along the road for eight miles.

The old water courses are called coulées. The road follows them, from the time it leaves Spokane Falls until it reaches the Columbia River at Ainsworth, for 150 miles. Timber is abundant east of Sprague, but not a tree is afterwards seen before the Columbia River is sighted, over 100 miles beyond. The coulées are rocky and desolate. There are stations all along, every few miles, and the company has planted shade trees at each of them, to show that, desert as this region appears, it only needs water and care to make the land productive.

Ten miles from Sprague the station *Harriston* is passed, and fourteen beyond the train reaches

Ritzville—(172 miles rom Heron.)—This is the starting point of what is likely to be a town of importance in the course of time. As yet immigration goes further north, and shuns the dry hills and plains below Sprague. The future will show that much of this soil is fertile, and well worth cultivation. Passing the stations *Lind* and *Twin Wells*, distant respectively seventeen and thirty-six miles from Ritzville, we arrive at

Palouse Junction—(217 miles from Heron.)—Here a branch railroad is in course of construction eastward, to develop the rich Palouse country, that is yet destitute of transportation. This road will bring to market the products of one of the most fertile and extensive agricultural regions on the Pacific Coast. The Palouse country extends from the base of the Cœur d'Alène Mountains westward sixty miles, so is partly in Idaho, and it reaches northwardly from Snake River seventy-five to 100 miles. The railroad will push east to the mountains nearly 100 miles, with branches to Moscow, Idaho, and Farmington, W. T. West of the Palouse there is very little arable land, but east of that stream is a fertile country of the best description. *Endicott* is a new town that the road will develop. It is in the midst of a farming country not yet settled and cultivated. The chief town in this region is *Colfax*, which has hitherto been the centre of business. *Moscow* is the terminus of the southeast branch of the road. It has the mountains for its eastern background, and a rich surrounding country that will build up its prosperity. The same is true of *Farmington*, twenty-five miles to the northward, to which another branch will extend. The Palouse region consists of a rolling surface of country, with rich soil even on the highest hill tops. There is no timber on its prairies, but abundance of it along the streams and on the mountains. The Falls of the Palouse River, distant a few miles from the Junction, are very beautiful. The next stations are *Lake* and *Eltopia*, distant from Palouse Junction ten and eighteen miles, respectively.

Ainsworth—(255 miles from Heron)—is at the confluence of the Columbia and Snake rivers, situated on a sandy plateau, with no attractions of green foliage or welcome shade. The great rivers join their waters and flow on. The town is built of temporary structures, and is uninteresting. Inhospitable as the surroundings are, a citizen of the place has proved, however, that its sage brush soil will produce well under efficient cultivation.

At Ainsworth the Northern Pacific Railroad Company is building a massive bridge across Snake River, which will be of great assistance to travel. At present a steam-boat, especially built for the purpose, ferries trains across.

Wallula Junction.—Fourteen miles below Ainsworth, and 214 miles from Portland, situated on the south bank of the Columbia, is the western terminus of the Pend d'Oreille Division of the Northern Pacific Railroad, and also the point at which the Oregon Railway and Navigation Company's road branches to Walla Walla. The Northern Pacific Railroad Company has erected a handsome building here for station and hotel purposes. Otherwise the place is as unattractive as Ainsworth. It is only important as a junction of roads, and as an eating station. The Walla Walla River comes in close by, but leaves all its fertile lands behind.

Pend d'Oreille Division.

At Wallula the journey to Portland, Oregon, proceeds over the line of the Oregon Railway and Navigation Company. This railroad connects Portland with the country south of Snake River, going eastward from Wallula to Walla Walla, and thence, still further eastward, to develop the country, touching Snake River at Riparia, where steam-boat navigation is maintained with Lewiston, in Idaho, the year round.

SKIRTING THE CLARK'S FORK.

Oregon Railway and Navigation Company.

WALLULA JUNCTION TO RIPARIA.—DISTANCE, 87 MILES.

Leaving the river at Wallula, the main line of the Oregon Railway and Navigation Company's railroad follows up the valley of the Walla Walla River, thirty-one miles to Walla Walla. The aspect of the country improves gradually as the distance from the river increases, and before reaching Walla Walla the country has become very fertile. The river is a small stream that pours into the Columbia without much demonstration—merely a channel cut through sand and sage brush, although further up there is an occasional fringe of willows. There is no appearance of even a village during this stretch of thirty miles, only side track stations, a few miles apart, for the transaction of the railroad business.

Whitman—(26 miles from Wallula and 5 miles from Walla Walla)—is merely a side track. It, however, marks the scene of a deplorable tragedy. In 1836 Dr. Marcus Whitman, a physician, who was also a clergyman, was sent out from the East as a missionary to the Cayuse and Umatilla Indians. Even at that early day Christian sympathy was drawn towards the aboriginal tribes of the Upper Columbia, and through this instrumentality the preservation of the Northern Pacific country to the United States is mainly due. Dr. Whitman established his mission at Wai-lat-pu, now Whitman's Station, where he faithfully labored among the red men. In 1847 he was making a professional visit to the Hudson Bay post at Wallula, from which his station was twenty-five miles inland, on the Walla Walla River, combining, in accordance with his usual custom, the practice of medicine with the preaching of the Gospel. When at Wallula, Whitman saw the arrival of a

Roman Catholic priest and his party, and heard the boast made that Oregon was certain to belong to the British, as Gov. Simpson, of the Hudson Bay Company, was in Washington, making negotiations to that end. This news weighed so heavily upon the missionary's mind, that, though late in the autumn, he prepared for and undertook a mid-winter journey across the continent, made representations to the Government as to the true value of the country, piloted the first wagon train through to the Columbia River the following spring, and so was greatly instrumental in preventing British ascendency in the Pacific Northwest. The year after Dr. Whitman returned to his mission, he, his wife and others, were massacred. It seems that the measles broke out among the Indians with great fatality. The medicine men of the tribes charged Whitman with causing the disease, and one night the cruel savages murdered their benefactor, with all his companions. The massacre occurred at the north end of the ridge, west of the railroad. There the victims were buried, and efforts are now making to raise a monument "to the memory of Dr. Marcus Whitman and his associate dead." This tragedy led to the Cayuse war of 1848.

Walla Walla—(31 miles from Wallula)—is beautifully situated upon an open plain that is watered by the divided flow of the Walla Walla River. Beyond it the Blue Mountains stand like a wall, and among the foot-hills is the richest agricultural district known. The city has 5,500 inhabitants, a handsome business street, with substantial blocks of stores—some very fine ones. Though no forest trees are native to the plain, the streets are lined with shade trees, usually poplar, and the gardens are filled with orchards and vineyards. The private residences are often beautiful. Near town is the military station of Fort Walla Walla, and the presence of troops adds something to the business as well as to the attractions of the city. There are several newspapers, banks, churches and excellent public schools. In the vicinity of the city is a large apple and peach orchard, which produces a lavish yield of excellent fruit. The importance of Walla Walla as a railroad point is increased by the fact that the Blue Mountain Branch will be extended to Pendleton. This branch follows the trend of the Blue Mountains, running parallel with them at ten or fifteen miles distance, opening a fine agricultural section. The road is already finished to *Blue Mountain Station*, twenty miles from Walla Walla, touching the pleasant town of *Milton*, nine miles distant from the present terminus.

Fifteen miles beyond Walla Walla the main road comes down into the valley of the Touchet River, which is a branch of the Walla Walla, and follows up that stream to Dayton. After passing several small stations,

Prescott—(51 miles from Wallula and 20 miles from Walla Walla)—is reached. Here is the end of a railroad division, where the company has established a round-house and shops. Five miles beyond is

Bolles Junction—(56 miles from Wallula)—where the branch to Dayton, thirteen miles long, deflects. The town of *Waitsburg* (population, 1,000), lies on this branch, and *Dayton* (a place of 1,600 inhabitants), is at the terminus of the road. Both towns do a large business. They are centres of trade for a rich agricultural region. The shipments of wheat are very heavy. In the Touchet Valley much attention is given to fruit growing, and the yield of apples, pears and peaches is quite large. The main line goes on thirty-one miles further to

Riparia—(87 miles from Wallula)—the present terminus on Snake River. There are no important points as yet between Bolles Junction and Riparia, as the country is, in a measure, undeveloped; but it is becoming rapidly populous and productive, and ships more wheat every year. The names of the small stations are *Menoken, Alto, Relief, Starbuck* and *Granger*.

Snake River Navigation.—At Riparia the railroad terminates for the present, and connection with the places on Snake River, as far as *Assotin*, which is near where the river emerges from the Blue Mountains, is made by means of the Oregon Railway and Navigation Company's steam-boats. These touch at landings on either shore, and bring to market heavy freights of grain and wool, grown in the lower Palouse country and between Snake River and the Blue Mountains. The chief points are *Penewawa, Almota, Wawawai* and *Alpowa* landings, where the business of the country is handled and its products shipped.

Snake River flows deep down in an immense cañon, whose cliffs are a thousand feet or more in height. Generally, the points are rock ribbed, for the strata show on every bluff. To ascend these cliffs is impossible, except some ravine is followed to its source, or a roadway is graded, carefully winding up the face of the acclivities. The shipment of grain would be attended with difficulty if the farmer had to haul his load down such tremendous hills, and spend hours returning to the plain above with his empty wagon. The evil is remedied by the construction of shutes leading for thousands of feet from the summit, down which the grain is poured to the warehouse on the river. There is communication between the various shipping points by means of a telephone, and the business is transacted with dispatch. The farmer simply delivers his wheat on the hill, and goes home rejoicing. The landing places named are merely warehouses, with perhaps a store, though goods are generally shipped and sent inland to Pomeroy, Pataha and a number of other towns situated in the farming region. The cañon of Snake River looks like an inferno, but the traveler who judges the country by this river scenery is entirely out of his reckoning. For example, to climb the grade opposite Lewiston is two hours' hard work, over two miles of distance, but when foot is placed on the surface of the rim rock, a rolling prairie region of excellent farming land is spread out far as the eye can reach. This is the case generally on the Columbia and Snake rivers.

Lewiston—(78 miles by steam-boat from Riparia; population, 1,000.)—This town was early created by the needs of the mining regions of middle Idaho. Mining was conducted with fabulous success in 1862, but the placers were exhausted long since. Now Lewiston is permanently supported by the agricultural and pastoral resources of a wide region. It is built at the junction of the Clearwater and Snake rivers, under the bluffs that are not high on the side of the river whereon it is situated. It has all the equipment for a thriving place. There are good hotels, two newspapers, a bank and heavy mercantile establishments, and its necessities will in time demand railroad facilities. At present the labor market of this whole region is taxed to push enterprises of great magnitude, links of thoroughfares that must be connected before the construction of branch roads can be successfully attempted.

Along the Columbia River.

[*Continued from page* 221.]

WALLULA JUNCTION TO PORTLAND.—DISTANCE, 214 MILES.

Returning to Wallula Junction, the journey westward is continued down the Columbia River upon the track of the Oregon Railway and Navigation Company. The immediate vicinity of the station is a wind-blown desert, with only a thread of green visible where the Walla Walla River struggles with the shifting sands to reach the Columbia. Northward rise the dark hills of Klickitat County, and just below the mouth of the Walla Walla the ridge that it follows ends in a rocky bluff. In cutting through this range of hills the Columbia has left two bold-faced and strata-marked headlands, facing each other, and affording the finest bit of landscape to be seen along the river for a hundred miles. Crags stand like ruins, more grand in their bronzed and rugged decay than any crumbling relics man has left. On one of them there are two colossal pillars, twin monuments of basalt, that can be seen in glimpses as the train passes. The popular legend is, that the line dividing Oregon and Washington runs between these pillars.

Passing these grand bluffs, the river courses through a region of low shores without a special object of interest for many miles. But, sterile and forbidding as this part of the route seems, still, not far from the road, on the Oregon shore, begin the rich farming lands of Umatilla County. Indeed, beyond the bluffs, on either side the river, there are arable lands within a few miles of the track. On the Washington side, a large fertile flat, containing thousands of acres, has been left vacant until quite recently, but a few settlers have now taken some of the land and made rude improvements. Irrigation is not very difficult where wind and water may be so easily utilized, and within a few years these wilderness shores of the Upper Columbia in many places will be made productive. The distance by rail from Wallula Junction to Dalles City is 127 miles. All the northern shore is that of Klickitat County, W. T., a region larger than two of the original thirteen States, and behind the low bluffs that bound the river for a stretch of sixty miles there are fine arable lands, upon which immigrants, doubtless, will shortly establish their homes. Passing the stations of *Cold Spring* and *Juniper*, the next halting place is

Umatilla Junction—(27 miles from Wallula)—a place that has been of commercial importance for over twenty years. It possesses little attractiveness, because there is nothing to relieve the monotony of sage bush and sand. The discovery of gold in the Blue Mountains, in 1862, made Umatilla the point for reshipment of goods. Merchandise intended for eastern Oregon and southern Idaho all came this way. There was at that date no Central Pacific Railroad, and the water transportation reduced the haul by wagons to the famous diggings of the Boisé Basin and the Owyhee about 300 miles. Umatilla then rose—it did not bloom—to be a place of many rough buildings and a fair share of rough trade. Long trains of patient pack mules, or impatient cayuse ponies, were going and coming, the picturesque array increased by the immense wagon caravans, often with eight mules, or horses, or as many oxen, as propelling power. That was a day of dust and weariness, but Umatilla throve upon it. Then came the decadence of the mines, the Central Pacific supplying what was left of the great mining camps of southern Idaho. But the dawning of the farming era inland, near by, soon built up thriving towns like Weston and Pendleton, and left Umatilla deserted until the construction of the railroad again brought it into prominence as a point of junction.

The Baker City Branch Line.—The Oregon Railway and Navigation Company has in view a system of railroads that is intended to develop all the agricultural areas tributary to the Columbia and lying south of Snake River. So far this system embraces the continuation of the trunk line from Wallula to Walla Walla and beyond, and a branch railroad to Baker City, in eastern Oregon.

The Baker City branch diverges from the main line at Umatilla, 187 miles from Portland, and is operated to *Pendleton*, a distance of forty-three miles, but is finished some distance beyond. The route from Pendleton lies over the Blue Mountains, crosses Snake River at the mouth of Burnt River, traverses the length of the beautiful Grande Ronde Valley, in which are the thriving towns of *La Grande* and *Union*, and passes by an easy divide into the Powder River Valley, terminating at *Baker City*. This place is already a business point of importance, commanding the trade of a wide section of eastern and southeastern Oregon. There are close by, in the Blue Mountains, a number of lucrative gold mines, and quartz lodes exist which will become valuable after the railroad is finished. At Baker City this branch line will connect with the Oregon Short Line, a part of the system of the Union Pacific Railroad, and this route will form a second line of direct communication between the Pacific Northwest and the Eastern States.

Pendleton is in the midst of an extensive and fertile farming region, that only waited for transportation facilities to assume first-class importance. The town reflects the value and progress of the surrounding country by its rapid growth and substantial prosperity. It will profit materially by becoming a railroad centre, the Oregon Railway and Navigation Company having already expended half a million dollars towards constructing a branch road direct thence to Walla Walla. Pendleton has two newspapers, a bank, several mills and factories, many well established business houses, good schools, churches and handsome private residences.

The Main Line Again.—After crossing the Umatilla River, here simply a sandy channel pouring a small stream into a large one, but tearing its way through the Blue Mountains, forty or fifty miles above, with the force of a powerful torrent, the stations *Stokes* and *Coyote* are passed—the latter unsentimental name derived from the slinking wolf of the country. Nine miles beyond the latter station is *Castle Rock*, standing between the track and the river, appearing like a Druidical monument or colossal altar of basalt. This rock is forty feet high, although a casual look gives no such impression. It is only noticeable as the single interesting feature in a scene of desolation. Perhaps it is a relic of the oldest superstition of the farthest West, with a wonderful history, if only there were any left to tell the reason that it stands so solitary. *Willow Creek*, nine miles further westward, is one of many beneficent streams which leave the mountains and fertilize and beautify the plains, and then lose themselves in the great Columbia. A bunch of green willows marks its exit and keeps its memory verdant.

Alkali—(73 miles from Wallula.)—This is a trading point, with a suggestive name, where has spread over the sandy hillside a street of rough board houses, that keep merchandise of all kinds, from a fine cambric needle to a Buckeye mower. This town of Alkali has turned its neighboring desolation into life and animation. There is a good farming country connected with it, which, not afraid of the name, builds up the fortunes of the traders. It looks like a business street, with a dozen to twenty stores and shops, and no private residences. When the farmers between Alkali and the Blue Mountains become more numerous, the town will respond with progress. Some July or August day, or some cold, clear winter morning, when the winds are all abroad, a flame will burst out in this pitch pine community, and when it expires for want of fuel the town-site will be in good shape for improved architecture. Such is the usual history of pitch pine villages. They generally undergo a fiery ordeal before they amount to much as business communities.

Blalock—(81 miles from Wallula.)—Blalock was named for an enterprising physician of Walla Walla, who devotes the income from his profession to farming projects. Dr. Blalock's farms, near Walla Walla, being very productive, he made up his mind that the country along the Columbia was equally fruitful. Now, with others, he has some thousands of acres in wheat on the bluffs that rise above the station.

The region of low shores and level country has here been passed, and the Columbia flows through deep-cut banks that are hundreds of feet in height. Desolation has become picturesque. Many rocky strata crop out on the overhanging cliffs, and reveal the processes by which nature wore a channel for the great river. Down in the cañon there is no pleasant shore, no fertile reach of valley land, no living green; no fringe of willows even waves along the bank. Instead, there are shifting sands that cover a great part of the land level with the river's flow. The town proprietors are trying to sow some suitable grass-seed on these sand reaches, in hope to rivet them by the aid of roots, and cure them of their restlessness. While great bluffs overhang the shores, on the south a ravine winds through the heights on an easy grade and climbs by three miles of good road to a rolling upland prairie, now partially under cultivation, and promising to become magnificent farming ground. As a specimen of western enterprise, it may be said that the "Blalock Wheat Growing Company" has been organized to redeem this land. Claiming the Government sections under homestead, pre-emption and timber culture laws, the stockholders have united their forces to construct a fence fourteen miles long to enclose the peninsula formed at the confluence of the Columbia and John Day rivers, and have 40,000 acres thus enclosed. Each man farms his own land, guarding any stock he may turn loose. The co-operation extends to the construction of this fence, the building of roads back towards the Blue Mountains, and the operating of a ferry across the Columbia at the town of Blalock. The company also owns the town-site, which is upon school lands purchased from the State. Common opinion condemned this region as unreliable for crops, as it had most other sections, until farmers went to work. It was necessary, in order to be successful, that cultivation should be attempted on a large scale. This organization was therefore formed, and it has succeeded in establishing the fact that good farming will redeem any part of these uplands and make them quite productive. This successful experiment will make Blalock a thriving point on the road, with a supporting trade.

The railroad has made this a division station, erected round-house and shops that will employ many hands and assist the agricultural interests in building up the place. The hills of Klickitat look on the river from the north, and there are farms planted on the bench lands that are not visible from below. From any of the grassy hill points of this region can be seen the colossal forms of Hood and Adams, whose snowy peaks are beacons of the "Inland Empire;" the long, blue line of the Cascade Mountains can also be traced to the west, and the dark crests of the Blue Mountains are much nearer, to the south. Passing *Quinn's*, seven miles from Blalock, we next come to

The John Day River—(97 miles from Wallula.)—This river enters the Columbia from the south, passing out through a walled cañon to reach the larger stream. A few miles distant its sunny shores are crowned with rich fruitage, for an enterprising farmer has thousands of bearing trees of various fruits. It is worthy of note that wherever the valley of any of these streams widens to admit of planting trees, and the land can be irrigated, the result is an excellent quality and prodigious yield. Down in these ravines the settler is sheltered from the vicissitudes of the seasons, and one wonders to see that the peach, apricot, almond, nectarine and finest varieties of the grape, including the raisin grape of California, all thrive, never knowing failure. The Snake and the Columbia rivers, and many of their tributaries, lying between the forty-second and the forty-ninth degrees of latitude, usually flow through deep cañons, and in their narrow valleys can be grown the fruits of California, as well as those native to the far North. There is proof of this when passing down below the John Day. The town of Columbus is built on a level reach on the Washington side, and fairly blooms with verdure. It is surrounded by peach orchards and other trees, a very oasis that gems the shores with a flush of green to cheer as the train flies on.

At the mouth of the John Day River the scenery along the Columbia improves in rugged grandeur, and the rapids in the river show tossing waters, the navigation of which requires a careful pilot. The railroad ignores the rapids, except when the unusual high water of exceptional years threatens the integrity of the track.

John Day was a man of early times, who seems to have had two streams named for him, the other John Day River entering Young's Bay, between Astoria and the ocean.

Geological.—It is noticeable that the rivers of western Washington and Oregon flow through natural valleys to reach the sea, while all the streams east of the Cascade Mountains have cut through deep cañons. The theory is that many centuries ago these eastern valleys were buried thousands of feet deep. Then, as now, the winds swept off the ocean from northwest and southwest, and when this region went through its different volcanic epochs, and fiery eruptions occurred, these winds swept the light ashes towards the east. Ashes and scoria and lava flow succeeded each other, covering deep down the lonely valleys, lakes and plains that existed when the mastodon roamed the earth. Prof. Condon, of the Oregon State University, learned years ago of the existence of fossil remains of the Pliocene period that had been found where the waters of John Day River, in cutting a channel, had exposed the bed of some old lake, now buried 1,500 feet. Other scientists, including Prof. Marsh, of Yale College, investigated in the same direction, and the treasures of scientific collections have been increased by remarkable specimens gathered in this John Day River region. Near the mouth of John Day River is a remarkable lava bed, over and through which the railroad passes. The space between the river and the bluffs is narrow, but is filled by black encrustations of lava, affording a glimpse of a region that, for a small extent, might be styled infernal. It is pleasant to know that on the heights above us are waving fields of grain, and that a little way up the John Day River is a bearing orchard of thousands of trees. Only thirteen miles from the mouth of the John Day River the Des Chutes is reached, another stream that heads far south. It collects the waters of the eastern shed of the Cascade Mountains for 200 miles, and sends a tributary to sweep up the streams that descend far south from the Blue Mountains. These two rivers are alike swift and turbulent, and come through deep worn cañons to join their floods with the Columbia. Crossing the Des Chutes, the road winds around its western bank to reach Celilo, on the Columbia, a wind driven spot, which, for a score or more of years, has been the western terminus of upper Columbia navigation.

Celilo—(114 miles from Wallula)—translated from the aboriginal, means "The Place of the Winds." The hills on the Washington side rise bluff and frowning. On the Oregon shore the shifting sands are freely driven by the unceasing winds. Above, for hundreds of miles, it is possible to send steam-boats up the Columbia and Snake rivers. But from Celilo to Dalles City, a distance of thirteen miles, navigation is forbidden by obstructions that are only overcome when it is necessary to take some steamer from the upper to the middle river. When the melting snows have swollen the Columbia to its fullest flow, and the waters boil so far above the rocks as to make the passage possible, then the coolest nerve is requisite and the most consummate skill called for.

In early days the corporation that controlled the river cemented its chain of transportation by constructing a railroad thirteen miles in length, from The Dalles to Celilo. It also had a shorter portage road around the Cascades, by which means the traffic of all the upper country for many a year was controlled. This portage has become a portion of the main trunk road up the Columbia River. Steam-boats sometimes load at Celilo to accommodate trade along the river, or to take freight up the Columbia above Ainsworth; but the glory of the river trade has departed. The fine steamers that used to navigate these waters have made the perilous passage over the Little Dalles, the Great Dalles and the Cascades, and are earning dividends on the Lower Columbia or Willamette, or else on the broader waves of Puget Sound.

Soon after leaving Celilo the classic regions of the great river are approached. If it is early summer, the hills to the north have not entirely thrown off their tinge of silver gray, given by the waving bunch grass. Later, after the grass has matured, these great hills, as well as the plains, turn to tints of golden brown. A short distance below Celilo the track curves around a steep basaltic cliff that overlooks the river, and wears the name of Cape Horn. Early travelers were not apt at names, and too often attached commonplace appellations to grand objects that deserve respectful treatment. This Cape Horn has no distinctive name, because there is another and grander Cape Horn on the river below.

The Little Dalles.—If it is early summer, and the Columbia is at flood, there will be seen below Celilo the Little Dalles of the river, a spot where the fall is enough to create foaming rapids for half a mile or more, as the pent up water rushes between the lava walls. The Little Dalles, however fine in itself, is rendered almost insignificant by comparison with the Great Dalles, six or eight miles below.

Indian Salmon Fishermen.—Over on the Washington side some Indians have their picturesque village—pole wigwams covered with mats or skins—to which distance lends all the enchantment. Here they come in the fishing season to catch salmon, which is dried for winter food. Half a century ago they came by thousands, and the desolate shores were alive with them. Every rock had its claimant, and every tribe its prerogative of fishing ground. Then there were no scores of canneries and packing establishments to devastate the fish, and no cunningly devised and cruel salmon wheel at the Cascades to swoop them up as they passed in myriads to the spawning grounds. Now only a few score Indians come to remind the whites that a remnant of the race still lives. If in June or July, a glimpse may be caught of some Siwash swinging his spear or wielding a scoop net over fierce rapids, waiting and watching for fish to ascend. The whole village is roused by the advent of the train, and, if at nightfall, the sons of the forest may be seen waving a greeting, their weird forms outlined against the sky. The family mansion, with its barking curs, and its smoking fire, is under the rocky wall. The Indian comes to the river for his fish supply as regularly as the year rolls round, and his cayuses browse near by on scant herbage found among the sage brush.

The Great Dalles of the Columbia.—During the months when the river is at low stage the Great Dalles is not a noticeable spot. It, however, well repays a careful examination, and rewards an observant visitor. It is five miles from the Great Dalles to Dalles City itself. All the way the scenery is inhospitable but surprising. To the west lies Dalles City, with its background of near hills and distant mountains. Towering above all, with its crown of snows, is Mount Hood, 11,000 feet high. The *tout ensemble* is magnificent. Glimpses of the city and the mountains high above it are to be caught as the train moves on, but they are apt to be neglected in watching the wonders of the river. When the flood is low, the Great Dalles affords a view of a wide expanse of lava encrustations with no river visible. You cross towards the north, climbing over the rough and rocky surface as you can, to find the river confined in a narrow cut close to the Washington shore. The flow is swift and dark. You pick up a stone—standing at the very brink—and easily throw it across from the Oregon to the Washington shore. And this is the mighty Columbia! You fling a second pebble so far that it surmounts the northern cliff, and might strike some animal grazing there. The truth is—and it is a wonder as well as a truth—at this place the tremendous volume of the greatest river on the west of North America is confined in a cut not much over sixty yards in width, but of fathomless depth. Frémont attempted, when he was earning his fame as an explorer, to measure the waters, but never could fathom them. The fact is undeniable that the river is turned on edge. You have seen the Great Dalles when the flood was low; but, if you see it when the river is full, you will find the wide expanse of rocks you clambered over to reach the chasm covered, many fathoms down, by a boiling flood that rushes furiously through every channel and hurls itself wickedly against huge rocks that bar its passage. For two miles or more the broad river is a furious torrent, that you cannot weary of looking at. For that distance the surface foams and rushes in a thousand fantastic shapes, boiling, where a hidden rock stands firm—a tremendous whirlpool, where there is room and depth for it. The foaming surges race and rush past one another, and sullenly disappear to give place to new shapes of frenzy. The stillest point is along the further shore, where the waters are deepest.

The United States Government caused a survey of the Dalles to be made in 1880, and the following description is taken from the notes of Mr. E. Hergesheimer, assistant in the United States Coast and Geodetic Survey, who performed the work:

"At this reach the river has worn through and carried away the successive layers of basalt for a depth of about 1,000 feet below the present summits, and, as the crests of the escarpments are still visible nearly to the summits, a fine opportunity is presented for the study of the type forms. The whole volume of the river here runs, for about one and a half miles, through a narrow gorge in the basalt, averaging about one hundred metres in width, and but sixty metres wide at its narrowest part. During the summer freshet it is much increased in volume, overflows its enclosing walls, follows and overflows some inferior parallel gorges, and is thus greatly increased in width. The water rushes and foams through the main gorge with great velocity, having a fall, at the time of our survey, of about twenty feet to the mile. The strata exposed average about seventy feet in thickness, and incline towards the ocean about one hundred and forty feet in a mile. They were all found to be distinct layers of basalt, except at a point on the southeast of the river, and seven hundred feet above the present level, where a deposit of lime is found, an interesting geological fact, historically.'

When the Hudson's Bay Company was in its prime there was no other authority in all the expanse of North America, from the waters of Hudson's Bay to those of Puget Sound, 4,000 miles apart. Communication between the far extremes was maintained by yearly journeys. In the high water season the *royageurs* of that company came sailing down the swollen stream, shooting the fearful rapids in their bateaux. Forty years ago American emigrants first essayed to make their way across the continent by land, and descended the river as they could. In the fall of 1843 the Applegate family arrived in Oregon. One of them tells of attempting the passage of The Dalles in a canoe, which was wrecked. He and another were saved, and a third was lost. His own experience was that he was sucked into a tremendous whirlpool, and rotated on its sides, looking up from the cylindrical depths to see the wrecked canoe whirling after him, and the stars shining clearly beyond. Some people do not find it convenient to believe all of this story, but those who know the narrator well recognize that it is very mildly told. It is true that Applegate was wrecked, taken into the whirlpool, and saved by stranding on a rock.

The weird aspect of nature at The Dalles, the black and rock bound shores, the river, in its always wild and sometimes fiercer moods, have no alleviation save the changing sky, that is almost always wreathed with smiles, and the lordly presence of Mount Hood, that wakes admiration. The beholder looks up from the sublimity of desolation around him, to see the same transformed into the ethereal and majestic, on a scale of grandeur that overawes while it impresses. Nature's moods are never trivial or wearisome on the Columbia. Below the tortured waters and the rock-ribbed shores that confine them, the Columbia broadens beautifully and becomes placid and inviting.

Dalles City—(127 miles from Wallula and 87 miles from Portland).—Dalles City is the eastern terminus of navigation on the Middle River. Here are still to be seen the fine boats of the Oregon Railway and Navigation Company, for navigation is maintained on this route the whole year round. The traveler may either remain in the car at Dalles City, or step off the train upon a steamer, and, with the current's aid, go down the placid stream while the train is coursing along the steep and wooded shores. Dalles City is one of the oldest settlements east of the Cascades, and is of considerable note, occupying the very gateway between the eastern and western divisions of the country.

The word Dalles, signifying swift waters, is applied as a general term to rapids on different points of the great river. The immigration of early times, as well as the *voyageurs* of the fur company, came to speak of The Dalles in general terms, and the word was finally applied to this locality as a specific designation. Missionaries tried in earliest times to establish a mission here, with limited success. A town sprang up as time developed the mines north and south, and agriculture and stock interests now support the thriving place. As the terminus of the Middle River Division, it received quite an income. It now lives on its actual surroundings, and will prosper more as development goes on. The town is built under a bluff, with farms and houses on the hill. The population is over 4,000. There are charming homes embowered among orchards and shade trees, several churches, good public schools, a fine academy and many industrial works, including the extensive shops of the railroad company. There are two newspapers, two large hotels, fine blocks of stores and pleasantly shaded streets.

From the city, but better still, from salient points on the adjoining hills, Mount Hood is seen grandly. From above the first bluff that terraces the heights behind the town, Mount Adams looks from beyond the Columbia, not equal to Hood, but still a mighty mountain.

The Dalles was even a noted place in the early days of settlement in this region. Here the emigrants, weary of the long march across the continent, and glad to avoid any further labor of road-making, leaving their empty wagons and tired teams to follow at leisure, themselves embarked with their effects upon rude boats to go down the Columbia to their destination at the Willamette settlements, over one hundred miles distant by the river. Reaching the Cascades, they made use of a portage six miles long at that unnavigable part of the great stream, resuming the boat journey beyond. This road around the Cascades is used to-day. There was not then, nor is there yet, a wagon road all the way down the Columbia, although an Indian trail at one time existed. In fact, there was no means of land communication between Dalles City and Portland until the railroad was opened in the autumn of 1882.

By Rail to Portland.—From Dalles City westward the railroad follows the river's edge, and the scene changes from treeless, desert looking shores to mountain views, that grow more interesting every mile. Soon after entering the mountains we find pines and firs scattered on the hillsides. Gradually forest growths increase. The mountains become at times densely wooded, and along the margin of the river maple, alder, ash and willow grow in tangled woods. The Columbia, from The Dalles to its exit from the mountains westward, has no valley. The mountains make the shores, leaving sometimes a fertile strip of bottom and occasionally some bench land. About twenty miles below the Dalles, Hood River comes in from the south and White Salmon from the north. Each has an arable valley near the Columbia. Save these two limited districts there is no farming land worth

Along the Cliffs of the Columbia.

notice for seventy-five miles. The track lies that distance through the great gorge the river has cut for its channel, working through a romantic region that has already become classic ground.

Hood River comes down from the snows of the great mountain, and has a charming valley, though not extensive. It has become attractive as a summer resort. Peaches thrive here, and many other fruits ripen to perfection. Hood River Valley, and that of White Salmon, on the north side of the Columbia, both have repute for their fruit-growing, and attract those who go to the mountains for summer rest and recreation.

Midway of the mountains are the *Upper Cascades* (169 miles from Wallula). Above, the mountains are beautiful and can be studied with careful attention. The placid river reflects the sky, and the heights are inverted with graphic effect in the limpid flow. Very beautiful views are caught as the road curves around projecting spurs of the ranges, by looking backward or forward across long watery reaches that have mountain and forest shores grouped in perfect beauty. Sometimes these views reach for many miles, to be shut off suddenly as the stream bends with the sweep of the mountains, when a new prospect is revealed. For placid beauty the upper river is supreme, although there are many really grand features of mountain scenery. Sometimes the forests climb to the summit. Sometimes they have been burned, and charred trunks stand against the sky. There have been fierce fires in these mountains, set by careless hunters, that have swept everything before them for many miles. Sometimes a pyramidal face of rock is seen, rising without tree or verdure of any kind, a mountain by itself. There is now and then a saw mill, or a shute from far in among the ranges, bringing down wood or lumber. There is scarce any other civilization visible.

Along the Cliffs.—Between the Dalles and Hood River there are two tunnels. The first is approached for half a mile directly under the face of a towering cliff that forms one of the most interesting objects on the route. This rough precipice has been blown away at its base, to enable the track to be laid. A rip-rap wall protects the bank, and the precipice almost overhangs it, ranging from 450 to 600 feet in height. Standing on the car platform, and looking up at this mighty wall, the beholder receives an impression that is likely always to remain.

During the building of the road a great deal of hazardous work was done along such places as this, sometimes at the expense of workmen's lives. The only way to blow off the face of this tremendous cliff was to lower men to the point at which they were to work with their tools, in rope slings. It was a risk that only men with the coolest nerve could undertake, and high pay was the inducement. Strength of muscle and power of will did not always save. The danger came by the loosening of rock above from friction of the rope. Fourteen men were killed by rocks falling on them in this way at different parts of the work. This road along the Columbia, of course, has easy grades, and so far compares favorably with the passage of the Central Pacific over the Sierra Nevada Mountains, in California. The same range runs north, thousands of miles, under the different names of the Andes, Cordilleras, Sierra Nevadas and Cascade mountains. The Columbia, having cut a gorge for itself, makes a somewhat tortuous and very expensive road, the great cost of building which can be understood by journeying over it.

Cascades and Mountains. Wooded ranges and abrupt cañons favor the existence of water-falls. Coming down from The Dalles by steamer, fine views are obtained of many famous falls or cascades on the Oregon side of the river. The railroad passes close to and almost under them, as they pour over the cliffs, sometimes so near that you are startled with the sound of plashing waters, and catch a glimpse of the foaming torrent as the train whirls by it.

The mountains become more abrupt as the "Heart of the Andes" is gained. Snows linger upon them, and hollows on the north side of them are filled with it until late in the summer. The snowy peaks that sentinel the range wear their kingly robes always, but wear them more lightly through the summer solstice. There will come a rainy time, sometimes in September, that will rehabilitate the mountains, fill the deep furrows and cover again the exposed ridges. After such a rain Mount Hood is suddenly transformed to a thing of wondrous beauty and purity.

The Sliding Mountain.—The Indians have a tradition that once the great snow mountains, Hood and Adams, stood close to the river at the Cascades, with a natural arch of stone bridging one to the other. The mountains quarreled, threw out stones, ashes and fire, and, in their anger with each other, demolished the arch. Before that time, the Indians say, their fathers had passed up and down beneath the arch in their canoes, and the stream was navigable; but when the arch fell, it choked the river, and created the rapids that now exist. The legend goes on to say that the "Sahullah Tyhee," or Great Spirit, was so angry with the contending mountains, that He hurled them north and south, where they stand to-day.

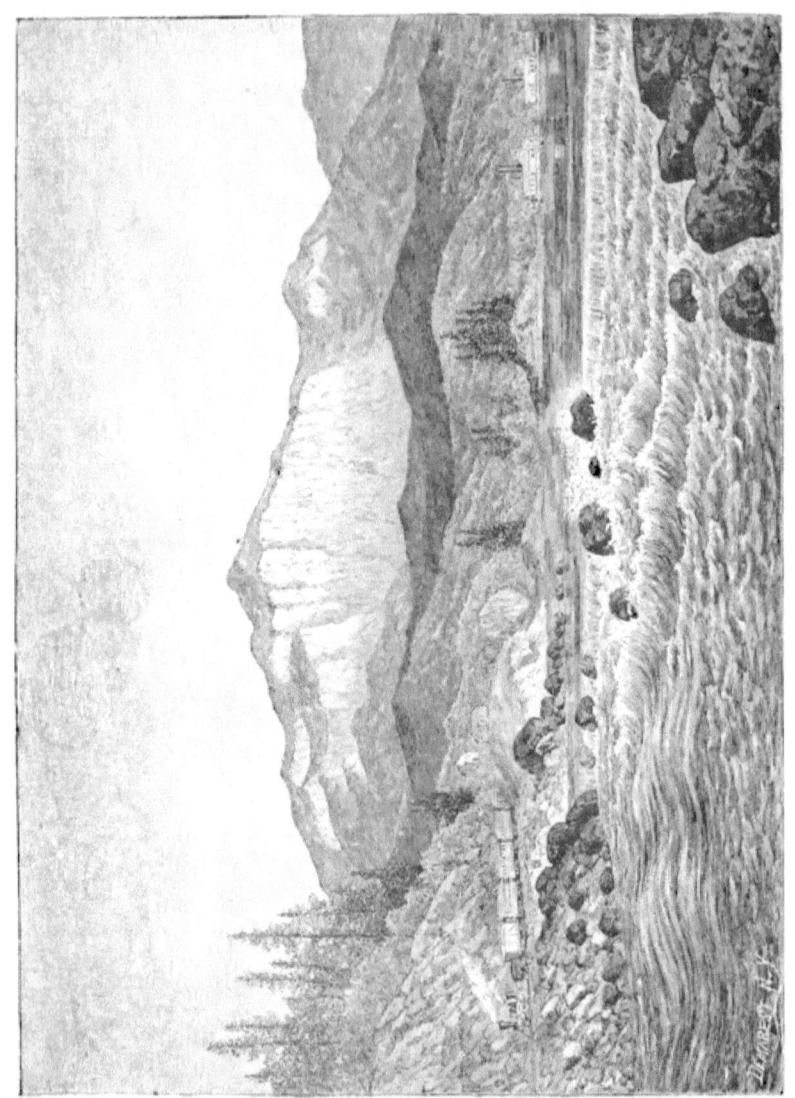

Cascades of the Columbia River.

This legend has some foundation, judging from the present conditions. It is evident, from the state of the shores and the submersion of forests, that some great convulsion has occurred and thrown down the rocky walls adjoining the river. Just above the Cascades the view includes beautiful islands, not far from the brink of the rapids, and between the islands and the rapids some ancient forest has been submerged, with the tree trunks still standing beneath the waves. It is commonly known to river men and steam-boat men that this submerged forest stands there, and it is often pointed out to travelers. How long since it grew on the shore no one knows. Indian legends are never accurate, and we can only surmise that it was long centuries before the white man came.

In connection with this legend, there are scientific data to establish the fact that some great convulsion has taken place and blocked the stream. When the rock walls fell and choked the channel, the effect was to raise the waters and deaden the flow for eight miles above. The work of engineers who have built and superintended the railways constructed around the Cascades for twenty years back, has demonstrated that, for a distance of three miles on the south, a great spur of the mountains is moving towards the river. The engineers who made the examinations connected with the canal and locks that Government is now constructing around the Cascades, have determined that the impending mountain of basalt rests on a bed of conglomerate, with substratum of sandstone, pitching towards the river. As the river wears away under the basalt, the rock masses move towards it. It is very possible that at some remote period, when the river had worn out a gorge, and precipices lined the shore, the waters undermined this wall and aided its descent on the incline of sandstone and conglomerate, so as to produce the effect which is seen, and confine to a short distance the fall that previously covered fifteen miles.

Mr. Theilson, chief engineer of the Northern Pacific Western Divisions, asserts that, a few years ago, when they were repairing the narrow gauge road, originally used for portage purposes, it was found that the track twisted out of line by the movement of the mountain. In one place it had moved seven or eight feet, and in other places ten. There was no mistaking the fact that there had been in two years a general movement of the whole mountain side for a distance of three miles. This testimony is conclusive, and it is very likely that the Indian tradition has its foundation in this fact. At the Upper Cascades the road goes close to the work carried on by the United States Government, of constructing a canal and locks around the rapids. This will require an outlay of millions, and it is done in the most substantial manner. It remains to be seen how the moving mountain will affect it.

The Cascades.—The Cascades are in about the centre of the mountain range. The river, that has flown so placidly all the way from The Dalles, has become wider, and spreads out in unbroken stillness, no motion being apparent. It is gathering itself for the plunge over the Cascades. In a moment it changes from a placid lake to swift rapids, that soon become a foaming torrent as the fall increases, and the waters encounter boulders in the stream.

Immediately at the Cascades the scenery is very fine. The mountains are grand, standing on the south like walls of adamant, and lifted to towering heights, their sides cleft open at intervals by deep ravines, the rock ledges of which are hidden by firs. Some of the rocky pinnacles and turrets along the heights are of strange, stern architecture.

On the north the mountains recede, and pyramidal forms contrast with tremendous frowning outlines, that stand like some Titanic fortress. There is a fine view of the Cascades from the train, and of the mountains on the north. At railroad speed the Lower Cascades are soon passed, and *Bonneville*, the point at which the steam-boats on the lower river make their landing, is reached.

The Old Block House.—Near the Upper Cascades, on the Washington side of the river, on a point of land that juts out so as to make a good defensive position, there is still standing an old block house, built thirty years ago, when the Indians were more numerous than peaceable. War broke out all along the coast, from British Columbia to California, in 1855. The Indians had some sort of unison, and outbreaks were almost simultaneous for that distance of 800 miles, though some of the more powerful tribes refused to join the alliance, and gave notice of danger. At that time the Cascades were already important as the portage where all things bound up the river had to make a transit. Suddenly the outbreak came. The block house became the refuge of all the settlers, who were defended by the male population and by a handful of soldiers, stationed there at the time under command of a young lieutenant named Sheridan. So the legend of the Indian and the wonders of nature are supplemented by a bit of history that has for its heroic character the now famous Gen. Phil. Sheridan, who was a favorite in the country thirty years ago.

A little way below the Cascades, on the south side, there is a canning establishment. Travelers feel much curiosity concerning great wheels that

Multuomah Falls, Columbia River.

float on the tide and revolve with the current. These are a cruel invention for taking fish. As they get into the rapids the salmon swim near the shore, and these wheels are placed in their way and take them bodily up without regard to size, landing them in great tanks for the use of the cannery. The quantity of fish sometimes taken in a few hours' good run is enormous.

Wonderful Scenery.—Soon after leaving Bonneville, a stretch of the grandest scenery of the grand river is entered. On the south, mountain summits stand like a wall, grouped at times like an amphitheatre, at other times assuming romantic shapes, and frequently affording views of falling waters that are very beautiful. Here is *Oneonta Fall*, 800 feet of sheet silver, a ribbon of mist waving in the wind. *Multnomah Fall* is double. The water plunges several hundred feet, gathers itself together and plunges again, about 800 feet in all. There are several other cascades of less note that never fail, and in early spring the face of the cliffs is threaded with them. A few miles westward are the *Pillars of Hercules*, two columns of rock several hundred feet in height, between which the train passes, as through a colossal portal, to the more open lands beyond. Near by is *Rooster Rock*, rising out of the river and pointing upward like a mighty index finger.

On the north side, a few miles below Lower Cascades, is *Castle Rock*, which rises by itself, no mountain adjoining, a thousand feet high. Castle Rock is a favorite view, and is well worthy of its reputation. Below, on the north side, is *Cape Horn*, a precipice over two hundred feet high, that rises abruptly from the water. This is another view that is much admired.

Every moment the tourist sees something to interest and attract. Attention is demanded in every direction, as new objects unfold. There are terraced heights, abrupt cliffs, crags in curious shapes, and mountain upon mountain to chain the eye. The unceasing panorama, with all its wonderful variety, is almost wearying, and a sense of relief is likely to be felt when the shores grow lower and the stream expands. Here the regions of western Oregon and Washington are reached. Islands are in the river grassed heavily and pastured by cattle. The shores, and the bluffs back of them, reveal homes and orchards. At last the Columbia has a valley, though not an extensive one.

Soon after leaving the mountain gorge, the track diverges from the river, and, passing through a forest region for about twenty miles, comes to East Portland, and then to Portland itself.

By River to Portland.

DISTANCE, 110 MILES.

Arriving at The Dalles, the traveler has the choice, as we have already said, of remaining on the train or of proceeding to Portland by steam-boat, the distance by water being 110 miles as against eighty-seven by rail. The river fleet of the Oregon Railway and Navigation Company is composed of first-class, speedy and commodious steamers, which are in every respect luxuriously equipped for the passenger service. The trip down the Columbia River is thoroughly enjoyable. From the deck of a steamer there is, of course, a far better opportunity to observe in detail the diversified beauties of the river than from the train. The scenery may be observed on both sides, and all the turns and changes of the stream are noticed.

At the Upper Cascades the steamer discharges her passengers, on the Washington side of the river, and here a short portage of six miles by railroad is made before reëmbarking on another steamer to pursue the journey to Portland.

The voyage onward, for a couple of hours, is upon the most romantic portion of the river. Castle Rock lies on the right hand, and on both sides, especially on the left shore, are to be seen foaming cascades pouring down the rugged faces of the mountains. Then comes Cape Horn, after which the river widens and the shores gradually become lower, with long, wooded islands in mid-stream, where dairy farming is carried on quite profitably.

Presently the city of *Vancouver*, W. T., is reached, and here the boat stops to take in fuel. The site of Vancouver is beautiful, and the place shows finely from the river. The east half of the city is devoted to the military, for this is the head-quarters of the Department of the Columbia. The storehouses, officers' quarters and barracks make an imposing appear-

Castle Rock, on the Columbia River.

ance. The shores on either side of the river, above and below Vancouver, are well cultivated and very attractive. From this point, looking west and south, Mount Hood (11,025 feet) is seen in perfect majesty. Twelve miles below, the steamer turns from the Columbia into the Willamette, and, looking north, other great mountains of Washington Territory loom up. St. Helens (9,750 feet) is sixty miles away, a vast white pyramid; Mount Adams (9,570 feet), seventy-five miles off, is partly hidden by the ranges; Mount Tacoma (14,360 feet), one hundred miles distant, on Puget Sound, or near it, is too remote to convey the correct impression of its grandeur, but can be plainly seen. There is one place, three miles up the Willamette, where five snow mountains can be seen at once on a clear day—St. Helen's, Tacoma, Adams, Hood and Jefferson, the last looking over the ranges for a long distance to the south. The panorama is magnificent, changing and opening at intervals as the steamer follows her course. Looking back, down the Willamette, a perfect picture is revealed where St. Helen's pyramid of white is framed in by the Willamette shores.

Approaching Portland by river, the traveler soon becomes aware that he is nearing a commercial city. River craft of all sorts and sizes, as well as ocean vessels, are found at the wharves of the city itself, one hundred and twenty-five miles from the ocean, representing the commerce of the world. East Indiamen, that have abandoned their former trade to steamers and the Suez Canal; ocean steamers, from the magnificent 3,000-ton passenger and freight steamships of the Oregon Railway and Navigation Company, to the business-looking colliers from Puget Sound, and the steam schooner that trades along the coast. These, and all sorts of river and coasting craft, are at Portland wharves. At the sight of them the fact is at once recognized that the journey across the continent is ended, and that the metropolis of the Pacific Northwest has been reached.

Portland—(214 miles from Wallula; population, 35,000)—is the commercial metropolis of the Pacific Northwest. It is already a beautiful city, and the centre of great wealth. It is finely situated on the west side of the Willamette River, twelve miles above its confluence with the Columbia. The population in 1870, including that of East Portland, was 1,103. This had swollen in 1880 to 23,000 souls, and in 1882 to about 35,000, and the ratio of increase in future is certain to be very much higher. The reasons for this are quite obvious. Portland's growth and progress are based upon the solid foundation of natural advantages of position. The site is so admirable that the limits of the city may be extended on every side. It is virtually a sea-port, to which large vessels may come direct from any part of the world, and find wharf accommodation. It lies in the very heart of a great producing country, which has no other outlet, and for which it must serve as a receiver and distributor of imports and exports. At no other point in the Pacific Northwest are these manifest superiorities offered.

Portland is the seat of a steamship company which runs lines of ocean steamers to San Francisco and Puget Sound, British Columbia and Alaska, as well as a fleet of river boats. It is also the greatest railway centre on the Pacific Coast—the grand terminus of a system which will completely develop the entire region.

The streets of Portland are regularly laid out, well paved and well lighted. The buildings of the business thoroughfares would do credit to any city, and the same may be said of many of the churches, the post-office, the custom-house and other public edifices, as well as private residences. The markets are good and spacious. There are public and other schools of various grades, a large library, well conducted newspapers, banks, commodious hotels, street cars, water, gas, manufacturing establishments, great wharfs and warehouses, telegraphic communication with all parts of the world, an immense wholesale and retail business, and, in fine, all the features of a flourishing modern city.

The wholesale trade of Portland in 1882 amounted to about $40,000,000, an aggregate probably never excelled by a city of similar size in the world, being an increase of twenty-eight per cent. in a single year. The value of building improvements in 1882 amounted to $2,977,000, of which sum $2,000,000 were expended upon business and manufacturing establishments alone. The factories of the city in 1882 turned out a product of $7,431,800, being an increase over the census returns of 1880 of $4,832,000. In 1882 the saw mills produced $1,000,000 worth of lumber; the product of the planing mills amounted to $400,000 in value, and that of the iron works to about $1,000,000. A great dry dock, capable of holding the largest vessels, has been built; and soon, through the enterprise of men of wealth, immense grain elevators, extensive flouring mills and enlarged iron works are to be erected.

Immediately opposite the city is the populous suburb of *East Portland*, adjoining which, on the north, is *Albina*, a rapidly growing place, where are situated the round-houses, car and machine shops, and other terminal facilities, of the Oregon Railway and Navigation Company, the Northern Pacific Railroad Company and the Oregon and California Railroad Company.

Oregon & California Railroad.

PORTLAND TO GLENDALE.—DISTANCE, 262 MILES.

The Oregon and California Railroad runs south from East Portland through the long line of valleys framed in by the Coast and Cascade ranges, and is in process of construction to the California line, where it will connect with the California and Oregon Railroad, a part of the Central Pacific system. Railroad building on this line was begun in 1868, and in 1870 the road had reached south to Roseburg, 197 miles.

Another branch was run, a few years since, from Portland, on the west side of the Willamette Valley, and finished to Corvallis, a distance of ninety-seven miles. This branch will be carried twenty-five miles further south to a junction with the East Side road.

Crossing the Willamette by a large ferry boat to *East Portland*, passengers find the south bound train waiting on the river side. East Portland spreads out over a gradually rising country that affords a favorable site for its growth. The train passes away from the river, seeking a natural route through a little valley, and in two miles reaches the engine and car shops of the company. Three miles beyond is *Wellsburg*, where a village has grown up around some factories. Seven miles from Portland is

Milwaukee, a small town lying between the railroad and the river, a mile away. There is here a large flouring mill, and orchards and nurseries surround the place. *Clackamas*, four miles beyond, is another station that supports a trading establishment and small village. After thirteen miles the Clackamas River, one of the swift flowing mountain tributaries of the Willamette, is reached, and crossed on a substantial bridge. Here a paper mill was established ten or twelve years ago, and this has become one of the permanent industries of Oregon

Oregon City—(16 miles from Portland; population, 2,000)—situated just below the beautiful Falls of the Willamette, is the oldest town in Oregon. A third of a century ago Oregon City was a thriving place, when Portland was an infant, and the trade of the country was once mainly transacted here. "The Falls" was the centre of civilization for this region, and the seat of its influence. Newspapers were published here in 1846, and the files of the little *Spectator*, which was printed on a hand press of primitive make, bought from Sandwich Island missionaries, afford interesting reading, as they tell of the trials, dangers and struggles of the pioneers to form a provisional government and obtain recognition from the Congress of the United States. Oregon City was the seat of the provisional government, instituted by self-respecting Americans when they could not induce Congress to take notice of their necessities and honest rights. The town lies partly in the cañon below the falls, in part on the open plain, and in part on the bluffs, which are reached by long flights of stairs that climb the cliff. The Willamette River, having been checked in its flow by a rock wall that crossed the valley at this point, has worn through the solid basalt, and left an abrupt fall of about thirty feet, that lies in the shape of a horseshoe. In making this plunge the river forever throws up a cloud of spray. This beautiful feature of the scenery changes with the changes of the river. With a full volume of water the falls are grand. It was this magnificent water power that attracted Dr McLaughlin, once chief factor of the Hudson's Bay Company, and induced him to become a citizen of the United States. The history of Oregon City includes many reminiscences of this fine old man, who, when the powerful company he represented ordered him to afford no aid to American immigrants, answered that his first duty was to be a man and a Christian. He did much to assist the early pioneers, and won small reward or appreciation from many of them for his generous deeds.

The Willamette Indians made much of the falls, as it was their favorite fishing ground. Their winter camps were here until white men came and monopolized the situation. Here the tribes fought battles for supremacy, and often met in high council. About the falls can be picked up many Indian relics, stone mortars and pestles, arrow heads and other articles.

There are already several manufacturing interests here, including a woolen factory that employs hundreds of people, and makes blankets, flannels, cloths, cassimeres, etc. This mill has been many years in successful operation. There are two flouring mills of large capacity, the product of which goes usually to Europe, and there are various other works that help to create a business for the place.

Pillars of Hercules and Rooster Rock on the Columbia River.

A good look at the falls may be had, as the train sweeps close by, following the shores and keeping under the bluffs for several miles. Four miles beyond is Rock Island, a rough place in the river, that steamboats cannot always pass. The shores are densely wooded with fir, cedar, maple, ash, alder and smaller growths. A few homes with orchards are seen, but no extent of fertile country is within view, and the traveler is apt to wonder where the Willamette Valley is, of which he has heard so much. The farms are on the bluff, and out of sight. The Tualatin and Yamhill rivers come in from the west, and the shores are rocky at times. Ten miles above Oregon City the river bank is left, and *New Era*, where there is a store, a mill and a post-office, is reached. Presently the train emerges upon *Barlow's Prairie*, a beautiful piece of farming land, with the swift and clear Molalla River on the north, and the sluggish and dark Pudding River on the south. Here, at last, is a farming country, that impresses with its fertility and beauty. Crossing Pudding River, the next station, twenty-nine miles from Portland, is

Aurora, a German settlement, until quite lately a colony under a sort of patriarchal rule. The death of its leader and founder has recently caused a winding up of its affairs and a division of property. Twenty-five years ago this was a heavily timbered ridge, valued so low as to be within reach of the colony, only a few members of which had means. All went to work under Dr. Kiel, their physician and manager, and the property was held in his name. Aurora soon became a favorite place. The colonists kept an hotel, cleared away the forests, made farms and planted orchards.

French Prairie.—The train is now on the northern limit of French Prairie, a rich farming region that was settled in early times. As far back as 1839 the Hudson's Bay Company found that constant hunting and trapping were exterminating the races of fur-bearing animals through the Pacific Northwest. This led to the introduction of farming, and the French Canadians and their half-breed children left the mountains for the plains, and began agriculture on this lovely prairie, which is not far from twenty miles square. The first field ever broken in Oregon is shown near the river, a few miles west of the railroad, where wheat was grown consecutively for half a century, almost without an interval of rest. The fact that this field retains its fertility, and produced thirty-five bushels of wheat to the acre in 1880, attests the lasting nature of Oregon soil.

The community that first occupied this prairie has in great part given place to Americans. It became too central and too civilized for the old mountaineers. The track of the Oregon and California Railroad is now marked by pleasant towns and villages, and fine farms are strewn over the whole fertile area. *Hubbard*, named after an early settler, is four miles beyond Aurora; *Woodburn* is a few miles further on; *Gervais* is in the centre of the prairie; *Brooks* is on the south edge of it. Gervais is named after the oldest inhabitant, a French Canadian, who was originally connected with the fur trade. The place has warehouses, stores, shops and a cluster of houses that give it a thriving appearance. The Roman Catholics have lately erected a handsome seminary building, and have also a church. The other places named are stations on the road and pleasant villages. The warehouses and stores at each point indicate the presence of a fertile country. As the train pushes forward through the Willamette Valley good farm houses are seen in all directions, surrounded by orchards, shade trees and gardens. In spring or summer these prairies are rich with waving grain fields, and the pastures are alive with flocks and herds of choice stock.

There are hills towards the Cascade Mountains, on the east, and across the Willamette, on the west. The hill country is as fertile as the prairies, and as well settled. The beautiful hill ranges of the Willamette, with their groves of fir and clustering white oaks, are charming features of the landscape. These hills are especially adapted to orchards, but they produce all other crops.

Salem—(53 miles south of Portland; population, 5,000)—is situated on the banks of the Willamette, which slope gradually back, so as to present a beautiful town-site. This was originally a mission station of the Methodist Episcopal Church, which built, in 1843, a large wooden "Institute," for the purpose of educating Indians. In this it failed; but it afterwards became a good school for white children, and resulted in the foundation of the "Willamette University," one of the best educational institutions of Oregon. Salem is the seat of the State Government. The Capitol will be an imposing building. The main structure is occupied, but the legislative hall is not finished, nor is the dome that is eventually to surmount the building. The State penitentiary and insane asylum are not far east of Salem, in a region of beautiful oak groves, plain but serviceable structures. Salem resembles a New England town of its size, with pleasant society and many beautiful residences. As an educational centre and the State capital, it draws together a refined and intelligent population. It covers a large area, has broad streets, good business quarters, quite a number of factories,

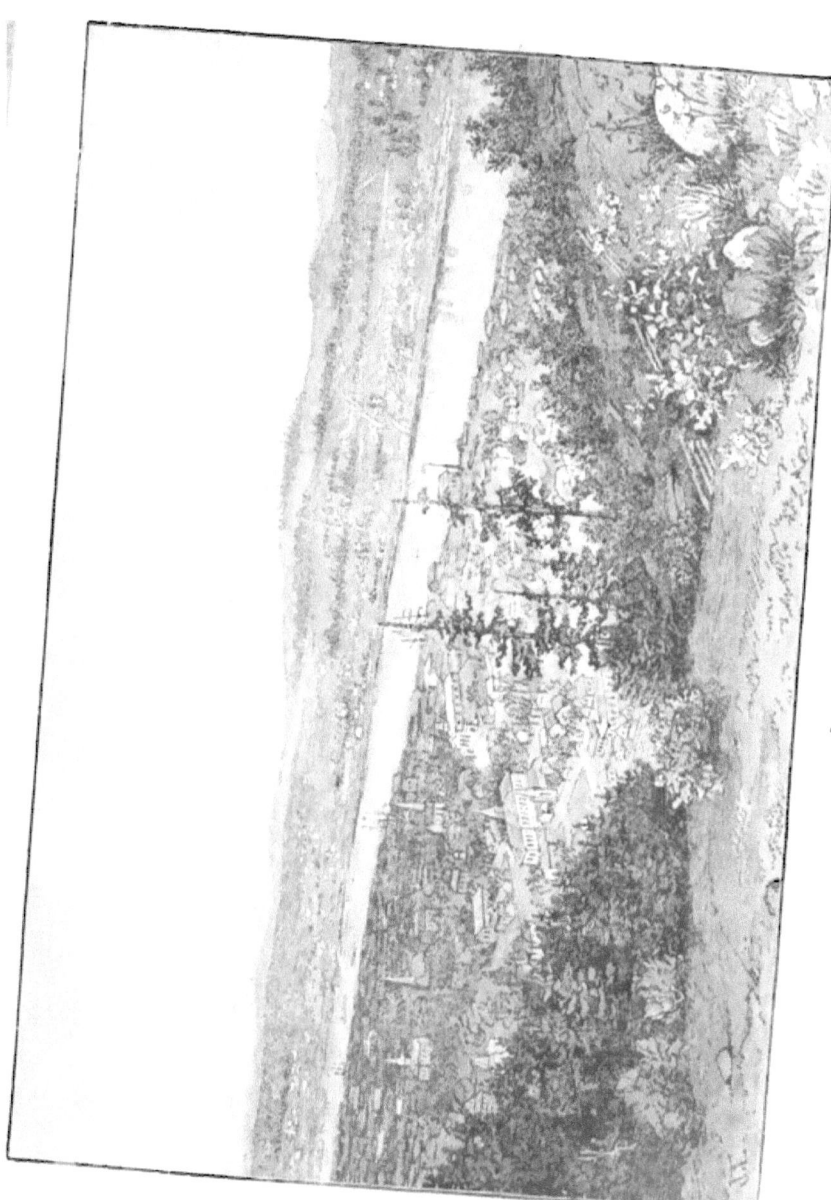

Portland, Oregon.

and possesses a water power, brought across country from the Santiam River, that favors its development into a manufacturing centre. There are two large flouring mills that manufacture for export. The first woolen factory on the Pacific coast was here, but burned down some years ago, and capitalists are now preparing to put up another. The town has various industries and several newspapers. A beautiful agricultural region surrounds it.

Beyond Salem the railroad skirts the hills, and passes through the narrow valley of Mill Creek to emerge upon the beautiful Santiam Prairie. It then sweeps around the base of the red hills for a dozen miles, passing *Turner*, a lively country town, with active flouring mills and considerable trade. Fourteen miles beyond Salem is *Marion*, a village close under the hill. On the Santiam River, seventy-two miles from Portland, is *Jefferson*, where there is a good flouring mill. The town has no special importance.

Crossing the Santiam, the train soon emerges from the bottoms upon the broad prairies of Linn County. This is especially the prairie county of the Willamette, and the road, lying along it for thirty-four miles, is almost a perfect level. There is a wonderful landscape. The whole valley, from the Cascades to the Coast Range, lies open, as it were, with outlying buttes along the base of the Cascade Range, relieving the descent from the mountains to the plains. Here the view of the snow mountains is superb.

Albany—(81 miles from Portland; population, 2,000)—is the principal business point of Linn County, and is situated on the river. Linn County takes the lead in agricultural products, and Albany is one of the thriving towns of the Willamette Valley and of the State. It has a collegiate institute, and good public schools are maintained. Albany also possesses a water supply that is brought by a ditch from the South Santiam across the prairie. It has several large merchant mills that do an extensive business. The people are enterprising, and encourage new industries.

Albany Junction is on the prairie, seven miles above Albany. At this point a branch road leads to *Lebanon*, an active town, with a good country surrounding it, eleven miles distant. Ten miles beyond Albany is *Halsey*, one of the busiest places on the Oregon and California Railroad. This is a point that ships largely of wheat, being the natural outlet for the rich prairie country east of it.

Harrisburg—(106 miles from Portland)—has the advantage of location on both river and railroad. This was a great point for shipment when steam-boats did the business of this region. The farmers of all the wide valley stored grain here, and made a large business for the town. Now the trade is divided up greatly, but Harrisburg still retains its fair share. The town has several hundred inhabitants, and has mill power that adds to its importance. Here the railroad crosses the Willamette River to the east side, continuing through prairies in Lane County for a long distance.

Four miles from the crossing the train enters *Junction*, named in anticipation of a union with the West Side road. This, it is hoped, will be consummated in 1883, when passengers can leave Portland by east or west side trains, as they may prefer. Junction is a town of importance, with extensive farming interests to maintain it. It has no water power to create permanent industries, but there are flouring mills in the vicinity. *Irving* is another shipping point, where grain is handled largely. It is also a prairie town, with a fertile region surrounding it.

Eugene—(124 miles from Portland; population, 1,600)—is the most important point in the upper valley. The wide prairies are left behind. The face of the country has changed to an undulating surface. The mountains are not far off, and some very striking buttes diversify the landscape. Eugene has a good trade from the country on all sides of it, and even from the valleys towards the ocean. It is beautifully situated, with buttes close by it, the landscape including a rich variety of scenery. The State University is here, represented by a handsome brick structure half a mile or so from the centre of the town. The Government has endowed this institution with a land grant, and the State affords it some aid. Private liberality has added to its resources, and it has every prospect of becoming the best regulated and best supported educational institution in the State. Eugene has many beautiful houses, and its share of business energy and active industries. Steam-boats used to come here at highest stage of water, but it was an irregular means of transportation, and since the railroad has been in operation, the traffic by river has been almost discontinued.

Beyond Eugene the road winds for twenty-five miles among the beautiful hills of Lane County, revealing charming landscapes at every turn. At last the Calipooia Mountains, a short range separating the Willamette and Umpqua valleys, are crossed by a low divide, from the summit of which, looking back, a magnificent view is presented of the broad, lowland country lying between the Cascades and the Coast Mountains. The Three Sis-

Iron Mountain, Cow Creek Cañon, Southern Oregon.

ters, of the Cascade Range, appear as snowy points on the horizon, and the majestic peaks of Adams, Jefferson and Hood have almost vanished in the distance among the clouds.

There are pleasant villages on the road beyond Eugene—*Springfield*, *Goshen*, *Cresswell* and *Latham*, each of which is a good point for trade. Descending from the Calipooia Mountains towards the Umpqua Valley, *Comstock's*, a station on the mountain side, is passed; and afterwards *Drain*, a centre of business situated at the edge of the beautiful Yoncalla Valley, is reached.

Oakland—(181 miles from Portland)—is a town of importance. The Umpqua Valley is a hill country, and the heights are crested with groves of white oak trees. Narrow valleys wind among these hills and offer romantic situations for homes. Sheep husbandry supersedes farming here, and fleeces of fine quality are exported. Oakland is among these hills, and has an academy that adds to its attractions.

Roseburg—(197 miles from Portland; population, 1,200)—is beautifully situated on the Umpqua River, and is the supply and distributing point for a wide region. Its growth has always been steady, and is likely to continue. It is the county-seat of Douglas County, and is supplied with churches, an academy, public schools, large business houses and a newspaper, with good milling power at the river.

Until 1882, Roseburg was the terminus of the railroad. At that time it was determined to extend the road to the California line, and there connect with the system of the Central Pacific, thus securing a through railroad route from Portland to San Francisco. The road is now in operation to *Glendale*, sixty-five miles beyond Roseburg, and a large constructing force is employed in pushing forward the work to its termination, 112 miles farther south.

After leaving Roseburg the road follows the course of the south Umpqua River along its well settled and picturesque valley, passing several unimportant stations, and halting at the pleasant settlements of *Myrtle Creek* and *Riddle's*, respectively twenty-three and twenty-nine miles south of Roseburg.

Engineering Difficulties. Owing to the roughness of the country beyond Riddle's, the construction of the railroad was carried on at great outlay of labor and money. The engineering difficulties which were encountered exceeded those of any work so far undertaken on the Pacific Coast.

To avoid Umpqua Mountain, over which the stage route formerly led, the road follows the windings of Cow Creek, a deep and romantic cañon, where much expensive tunneling has been done. This cañon is so tortuous that it was necessary to build not less than thirty-five miles of railroad in order to get over one-third of actual distance in a direct line from point to point. The cost of constructing the road on this stretch was not less than $40,000 per mile; or, at the rate of $120,000 per mile if the actual distance between the entrance of Cow Creek cañon to the point where the railroad leaves it, at Wolf Creek, is considered. In pursuing the sinuous length of the cañon the road, in a distance of twenty-seven miles, passes through no less than eight tunnels, the aggregate length of which exceeds 5,000 feet. The scenery is very attractive, not alone here, but all along the line, and this feature will make the Oregon and California road a favorite route for pleasure travel.

The extension of this railroad has already opened to traffic the Rogue River Valley, a portion of Oregon possessing not only great advantages of soil and climate, but a wealth of mineral and forest lands. The name Rogue River is a corruption of the designation which was given to the stream by the trappers of the Hudson's Bay Company. They called it Rivière Rouge, owing to the banks of red clay which exist toward its mouth, the washings of which sometimes reddened its water. The Rogue River Valley is little inferior in size to the Willamette, and is so sheltered and fertile as to readily produce everything that can be successfully cultivated in northern California. Especially is it adapted to vineyards and peach orchards. Jacksonville and Ashland are the principal towns, each with about 1,000 inhabitants.

There are mines of cinnabar, nickel, copper, iron, lead, silver, gold and very good coal through all the region penetrated by the railroad, the lack of means of transportation alone having hitherto prevented the full development of this natural wealth. Gold was discovered in the Rogue River Valley as early as 1851, and the placers proving to be rich, the town of Jacksonville was soon founded. The Indian tribes of southern Oregon, however, a braver and wilder race than any other with whom the whites came in contact, disputed the presence of the miners, and conflicts followed. In 1853 the red men avenged their grievances by applying the torch to every settler's house and using the scalping-knife upon women and children. This was a terrible event in the annals of settlement, and it was long before the savages were subdued. The remnant of the Rogue River Indians now occupy the coast reservations opposite the Willamette Valley, and engage in civilized pursuits. Sam, their chief, was considered too dangerous a character to be left permanently with his tribe, even after they were exiled from their lands. He often incited his warriors to break away from the reservation and return to their old hunting-grounds, whence they were brought back at great cost of time and trouble by the military authorities. At last it was determined that the rebellious chief and a few of his more troublesome companions should be confined in California. Passage was secured for them upon one of the steamships which ply between Portland and San Francisco, but the prowess of the prisoners was underestimated, and one day they almost got possession of the ship, their mutiny being quelled with difficulty. But the traveler who sees the beautiful valley these Indians once inhabited, surrounded by hills and mountains, and abounding with game, is not likely to wonder at their remarkable attachment for the homes and the hunting-grounds they and their fathers owned until the day that the white man, in his search for gold, dispossessed them.

Owing to its isolation, southern Oregon has been hitherto little known, and it has been less visited than any other portion of the State west of the Cascade Mountains. Its mild climate and fertile soil, and its generous resources in mines and forests, will cause it to respond quickly to the transportation facilities which finally have reached it.

West Side Division.

PORTLAND TO CORVALLIS.—DISTANCE, 97 MILES.

In addition to the railroad from East Portland through the Willamette, Umpqua and Rogue River valleys, the Oregon and California Railroad Company own and operate a second railroad, familiarly known as the "West side road," which runs from Portland to Corvallis, a distance of ninety-seven miles, and is to be extended to Junction City, in Lane County, twenty-three miles farther, there to connect with the main line of the same system.

The West Side Railroad, after crossing the hills behind Portland, passes westward through an almost unbroken forest region, until it reaches the plains of Washington County. These plains consist of prairie openings, surrounded by wooded belts, and are commonly known as the Tualatin Plains, from the river of that name, which empties into the Willamette, above the falls of the latter stream. *Summit* is the first station, reached by a grade cut on the mountain side for four miles. Here an opening in the hills permits the road to pass through to the plains. There is no town at Summit, but close by are flourishing prune orchards several hundred acres in extent. This hill and forest land is naturally adapted to fruit culture, if the situations are not too much exposed. Eleven miles out is *Beaverton*, a good point, named from the existence of a large beaver marsh, the soil of which is famous for the growth of onions and other vegetables. This land sells for $300 an acre, and, when it is under cultivation, proves immensely productive.

Reedsville—(16 miles from Portland)—named after Mr. S. G. Reed, who, with Mr. W. S. Ladd, a Portland banker, has interested himself in importing and breeding fine stock. The Reedsville farm is admirably kept, and is prepared for raising and training horses, especially good trotting animals. The short horn herd, owned by these gentlemen, is of the best stock known. This stock farm is an important feature of Washington County.

Hillsboro—(21 miles from Portland)—is the county-seat. It supports a newspaper, and is a thriving town, with a large trade, in the midst of a good farming country that was settled and cultivated at an early day. *Cornelius* and *Gaston* are two other business points in the same county.

Forest Grove—(26 miles from Portland)—is distinguished as an educational point. It is the seat of Pacific University, an institution that has preparatory and academic branches. This college is under the management of the Congregational Church, but is not sectarian in character. It has, for a new country, a large endowment fund, and is prospering steadily. Forest Grove is also the seat of an Indian Training School, for both sexes, under Government control. The pupils are received from different tribes in Alaska, and on Puget Sound, and along the Coast, as well as from the interior. The boys, in cadet uniform, and under military discipline, make a fine appearance. The girls are improved by giving them a knowledge of domestic duties, as well as of the common branches of learning. This school is a decided success, inasmuch as it proves how much education can do for savage tribes. The object is to fit all the pupils for the pursuits of civilized life, and for an appreciation of domestic ties.

Leaving Washington County, we enter Yamhill, considered the choicest region, if any choice is possible, in the Willamette Valley. Yamhill County has generally a rolling prairie surface, extending from the Willamette River to the Coast Mountains, twenty to thirty miles wide. The soil is excellent, and Nature has favored the county with running streams and native forests, giving every facility for the establishment of homes. This, indeed, is the case through the entire valley, with its 4,000,000 acres and its nine rich and prosperous counties. *North Yamhill*, thirty-nine miles from Portland, is a prosperous town, but

McMinnville, a few miles beyond, is the best place for business in the county. It has mills, factories, stores, hotels and a newspaper, and reflects the prosperity of a rich agricultural community. *Amity* is also a pleasant town on the open prairie, and central to an excellent farming district. Passing McMinnville, there are several new towns along the road, but no place of note, until

Independence (76 miles from Portland)—in Polk County, is reached. Here, for the first time, the road touches the Willamette River. Independence is a lively business point, and has grown very rapidly since the advent of the railroad in 1881. A few miles inland from Independence is *Monmouth*, on the open prairie. This place is the seat of a Christian College, conducted by the denomination commonly known as "Campbellites," of whom there are many in Oregon. It is a prosperous institution, and has a large patronage. There are a number of small towns on the succeeding twenty miles of the railroad, but no place of special importance is reached, until we come to the present terminus at

Corvallis—(97 miles from Portland.)—This town is situated immediately on the west bank of the Willamette River, in Benton County. The country behind it is level prairie for some distance, but changes to upland. Mary's Peak, the highest point on the Coast range, is in plain view, and the scenery from the town is very fine. Corvallis will always be one of the best places in Western Oregon. It supports two newspapers, has its share of mills and warehouses, and its merchants do a large trade with the surrounding country. The State Agricultural College is established at Corvallis, and the advantages afforded by this institution add to the growth and prosperity of the town. Corvallis is opposite one of the best passes of the Coast range, leading to Yaquina Bay. A wagon road, favored by a land grant, has made connection with the sea coast easy, and some work has been done towards constructing a railroad to connect Corvallis with the bay.

The work of building the West Side Railroad to its junction with the East Side division of the Oregon and California Railroad will be carried on at an early day, and then each side of the Willamette Valley will be connected with California as directly as possible. There are no towns of note on the unfinished portion of the route, and no features of the country to be traversed call for especial notice.

The Oregonian Railway Company.

The Willamette Valley is 125 miles in length and from thirty to fifty miles in width. Nature has blessed this valley with a navigable river, and this method of travel and traffic has been so lavishly supplemented by railroad enterprise that probably no other area of similar size and population within the bounds of the United States is so well supplied with transportation facilities.

Beside the Oregon and California railroad lines, on the east and west sides of the Willamette River, there is still another railroad system called the Oregonian Railway, which traverses the most fertile lands of the valley. This Oregonian Railway is a narrow gauge line, which was built by a company of Scotch capitalists a few years since, and was afterwards leased by its owners to the Oregon Railway and Navigation Company. This railroad runs out eastward and westward from the Willamette River, thirty-two miles from Portland, crosses the tracks of the Oregon and California Railroad, and then pursues a southerly course on the flanks of these standard gauge lines, passing between them and the Cascade Mountains on the east and the Coast range on the west. It is brought into direct relations with Portland by the steam-boats of the Oregon Railway and Navigation Company on the Willamette River.

The East Side Division—(80 miles long)—of the Oregonian Railway crosses French Prairie from *St. Paul's*, in Marion County, to *Woodburn*, eight miles; thence runs ten miles to *Silverton*, a lively trading place and milling point on Silver Creek, close to the Waldo Hills. Following around the base of the Waldo Hills the road passes the upper end of *Howell Prairie*, one of the richest spots in the Willamette Valley, and keeps on close to the hills, within six miles of Salem, passing *Aumsville* and the Santiam River without touching any towns of note, running almost forty miles in Marion County.

After passing the Santiam the route is south, a little west of the flourishing town of *Scio*, in the forks of the Willamette, then across the south Santiam and east of *Lebanon*, a good town in the eastern part of Linn County, touching *Brownsville*, one of the lively towns of the same county, distant 100 miles from Portland, and situated at the edge of the hills overlooking the broad Linn County prairies. It then runs along the east side of Linn County, with foot-hills and outlying buttes in the foreground and the Cascade range back of all. This is a beautiful stretch of country to ride over, especially when the yellow harvest is ripe. Across the wide valley, to the west, is the Coast range, with Mary's Peak prominent. To the southward are the Calipooia Mountains, that tie the great lateral ranges together at the head of the valley.

This East Side division terminates at *Coburg*, in Lane County, having traversed Linn County over forty miles.

The West Side Division—(44 miles long)—begins at *Dundee*, on the Willamette River, thirty-two miles from Portland, passes *Dayton* and *Lafayette*, old towns in Yamhill County, crosses the standard gauge near *McMinnville*, and then explores the west side of Yamhill and Polk counties for thirty-seven miles, throwing out a branch seven miles in length to catch the trade of the rich farming country around *Sheridan*, close under the Coast range. It visits *Dallas*, the county-seat of Polk County, passing first through *Perrydale*, a prairie village, that ships largely of home products. The route lies through *Monmouth*, in the very heart of Polk County, and ends at the new town of *Airlie*, in Benton County, named after the Earl of Airlie, who was largely interested in this railroad enterprise, and who died suddenly in Colorado.

A Trip on the Willamette River.

The Willamette River during many years was the avenue of all the traffic of western Oregon, and daily lines of steam-boats waked the shores to activity. River towns flourished, and landings that had little trade beside shipped large quantities of grain and other produce. Boats ran all summer to Salem, and to Dayton on the Yamhill River, and a great part of the year to Albany and Corvallis. Harrisburg, where the railroad crosses the Willamette, was formerly a great shipping point, even through the late fall and winter season. At the highest flood, boats ascended to Eugene and Springfield. These were occasions which tried the qualities of river pilots, and certain captains were famous for their proficiency in knowing the bars and snags of the Willamette. Sometimes, too, the chief tributaries of the river could be ascended. But constant skill and care were required in conducting the commerce of the Willamette Valley under these conditions.

At this date there was scarcely any commerce on the upper and lower Columbia or on Puget Sound. The travel on the upper Columbia was chiefly confined to the voyage down of the immigrants who had "crossed the plains." The traffic up the river was mainly in conveying Government troops and supplies. The lower river had no importance that demanded good accommodations. There were no canneries and few fisheries, and Astoria had no foreign commerce to feed it. The Willamette River was virtually called on to transport the bulk of the passengers and freight of the whole region. The consequence was that boats were built above and below the Falls of the Willamette, and a horse tramway was constructed, a mile in length, to connect travel at this point of interruption. This state of affairs called the People's Transportation Company into being, and Oregon enterprise culminated in the construction by that organization of a breakwater and basin and other improvements at the falls which gave it the monopoly of navigating the river. Subsequently, as people came into the country, another organization, known as the Oregon Steam Navigation Company, was formed, for the purpose of profiting by the growing traffic on the Columbia River, and for a long period these two corporations amicably divided, without any competition, all the commerce of the Pacific Northwest.

Finally, railroads came to the relief of western Oregon, simplifying and systematizing its business relations. This caused towns to spring up along the railroad, and those river towns which were not already railroad towns lost their prosperity. There is, however, still a steady commerce on the river to supply districts that are not reached by rail.

The boats of the Oregon Railway and Navigation Company leave Portland daily for the upper Willamette and Yamhill rivers, and bring down a great deal of grain, stock and produce, as well as passengers who live near the river. The banks of the Willamette are somewhat uninteresting. They are often covered with ash, alder, maple and cottonwood forests, and few residences or improvements are observed from the passing boats. Some towns are quite near, but usually so hidden by the trees that the visible portions are only mills and uninteresting warehouses. Oregon City, Salem, Independence, Buena Vista, Albany, Corvallis, Harrisburg and Eugene are all on the river, but their best points are lost to sight.

The Falls of the Willamette are very beautiful. On the west side they are surmounted by means of a canal and expensive locks, excavated through solid rock at a cost of five hundred thousand dollars. This enterprise was undertaken with private capital, aided by an appropriation of two hundred thousand dollars by the State. The property eventually went into the possession of the Oregon Steam Navigation Company, with the steam-boats and other plant of the old People's Transportation Company, and the use of it was enjoyed by them until a few years since, when it was turned over to the Oregon Railway and Navigation Company.

It is well worth while, if the tourist is able to spare the time, to go up the Willamette by boat, if only to enjoy the passage of the locks, with the accompanying view of the falls and of Oregon City, part of which place is built on the bluff, while the business street, with its mills, factories and warehouses, lines the river, and can be seen to advantage when passing the locks.

Northern Pacific Railroad.

(PACIFIC DIVISION.)

PORTLAND TO TACOMA.—DISTANCE, 147 MILES.

The Pacific Division of the Northern Pacific Railroad has been in operation since 1874, from Kalama, on the Columbia River, to Tacoma, on Puget Sound. Hitherto the distance of thirty-eight miles between Kalama and Portland by river has been covered by means of the steam-boats of the Oregon Railway and Navigation Company. Now, however, the railroad is in process of extension from a point opposite Kalama, on the south side of the Columbia River, to Portland, and it is believed that through connection by rail will be made during the autumn of 1883. This link of forty-two miles by rail is all that is required to perfect the direct transcontinental railroad between the East and Puget Sound, opening the great northern route from ocean to ocean.

The road from Portland follows for about eight miles the west shore of the Willamette River, and reaches the head of Sauvie's Island. Thence it continues down the west arm of the river to the point at which the stream empties into the Columbia at St. Helen's. Sauvie's Island is nearly twenty miles in length. It is a rich piece of land in the delta, formed by the junction of the rivers. There are no towns of note for all this distance. *St. Helen's* is a thriving place, with a large lumber mill, and with expectations of more rapid growth, based upon the fact that it has some good agricultural country immediately around it, and that the Nehalem region, from which it is separated by the hills, is becoming populated so fast that the trade of the farmers must centre at this point. The train runs on still farther, to *Columbia City*, another expectant locality. There is little population in the town, and very little along the shores of the Columbia itself.

Connection with Kalama, on the Washington side, will be made by means of an immense ferry boat that can take a whole train across the river at once. *Kalama* was never a place of note, its importance resting solely on the fact that the railroad terminated here.

Leaving Kalama, the track follows down the Columbia a few miles, then turns up the valley of the Cowlitz River, and strikes across the country for Puget Sound.

In the Cowlitz Valley are rich bottom lands that were taken by settlers at a very early day. These farms are famous for producing hay and dairy products. For 105 miles, from Kalama to Tacoma, the route is through a wooded region that is, for the most part, sparsely populated. The soil of all this area is fertile, but the thick forests are likely to deter settlement to a large extent, until the vast expanses of open and productive country east of the Cascade Mountains are fully occupied and brought under cultivation.

The stations passed on the way are *Carroll's*, *Monticello*, *Cowlitz* and *Olequa*, distant, respectively, five, eight, eleven and twenty-eight miles from Kalama.

Passing out of the Cowlitz Valley, the road reaches the Chehalis River. This stream runs in a northwest direction, and empties into the ocean at Gray's Harbor. Its valley, varying in breadth from fifteen to fifty miles, is the largest and most valuable agricultural region in Western Washington. The stations *Winlock*, *Newaukum* and *Shookumchuck*, respectively distant from Kalama thirty-seven, forty-eight and sixty miles, are left behind, and the broad valley of the Nisqually River is reached.

A fine view of Mount Adams, away to the eastward, on the further side of the Cascade range, is to be obtained at several points, as the train goes northward. It is seen across the wooded valley of the Nisqually, its white mass in bold relief against the sky, its sides seamed in summer with outcropping rock ridges, the hollows between being filled with never-melting snows.

Tenino—(66 miles from Kalama.)—The Olympia and Chehalis Valley Railroad, a narrow gauge line, fifteen miles long, owned by an independent corporation, connects *Olympia*, the capital of Washington Territory, and the county-seat of Thurston County, with the track of the Northern Pacific Railroad at Tenino. The road passes through a dense forest, touching the stations *Gillmore*, *Spurlock*, *Plum*, *Bush Prairie* and *Tumwater*, the latter a lively manufacturing village, with fine water power, on the outskirts of Olympia. The capital is delightfully situated at the head

Tacoma, W. T.

of Hood's Canal. It has now about 2,000 inhabitants, and is constantly growing in population and wealth. All the United States officials in the Territory are stationed here, and give tone to the society of the city. The streets are broad and well shaded, and the principal thoroughfare is lined with stores. Churches, schools, a newspaper, a bank and several hotels are among the features of the place. Thurston County, in which Olympia is situated, is densely wooded, and lumbering is a leading industry. There is a great extent of prairie and bottom land in the county, adapted to stock raising and mixed farming.

The Northern Pacific Railroad deflects eastward after leaving Tenino, and at *Yelm Prairie*, fourteen miles beyond, there is a revelation of unsurpassed grandeur, provided the sky be cloudless, in the view of Mount Tacoma, the loftiest of all the snow mountains. As the train rushes onward, occasional breaks in the forest allow the sight of this snow-clad peak to great advantage. It is about forty miles distant, although its vast bulk is so distinct that it seems much nearer than that. The road comes out of the forest on the gravelly plains of the Nisqually, and for twenty-five miles passes over this stretch of valley land until it stops at Tacoma.

At *Lake View*—(96 miles from Kalama)—in the Nisqually Valley, the last station before Tacoma is reached, there is a group of beautiful lakes surrounded by pine groves and stocked with fish. Some gentlemen have chosen this spot for a summer resort, and built cottages around Gravelly Lake, one of the most attractive of the cluster. This is considered one of the pleasantest summering places on the north coast. There are fine drives over the level prairie, and the ever-changing views of Mount Tacoma are magnificent.

Tacoma—(105 miles from Kalama; population, 3,500)—was the first point touched by the Northern Pacific Railroad on the waters of the Pacific Ocean. It occupies a commanding position, and has an excellent harbor, capable of receiving the largest ocean-going vessels, which are loaded at the wharves with coal, lumber and other productions of the region. Commencement Bay, on the east, opens upon the fertile valley of Puyallup, and beyond, in the near distance, rises the grand form of snow-covered Mount Tacoma. The railroad reaches the shores of Commencement Bay, under a steep bluff, the summit of which is crowned with pleasant residences. Views from the bluff are superb, including the Olympic Mountains, on the peninsula between the Sound and the Ocean; the full sweep of the waters of the inlets near by, the wide expanse of Commencement Bay, and the grand mountain scenery beyond.

The growth of Tacoma has been very rapid. In 1880 the population was less than one thousand, and at this time (the summer of 1883) the estimated number of inhabitants is 3,500. Business has kept pace with population. Railroad shops and boiler works give employment to many hands, and the various industrial establishments of the city are constantly increasing their capacity. About 250 houses were erected in 1882 and building operations continue active, the demand for stores and dwellings being so great that tenants usually engage to take the new buildings before they are ready for occupancy. The chief business street is solidly built up, and many of the stores do a large wholesale trade. Much good farming land is tributary to Tacoma, between the Puyallup and Nisqually rivers, extending to the Cascade Mountains, including the rich lands of the Puyallup Valley, which are noted especially for producing heavy crops of hops. Tacoma is well supplied with schools, newspapers and churches, among the latter being a beautiful Episcopal church, built of the fine, blue stone of Whatcom County, by C. B. Wright, Esq., of Philadelphia, as a memorial.

Cascade Branch.

TACOMA TO CARBONADO.—DISTANCE, 34 MILES.

The Cascade Division of the Northern Pacific Railroad is in operation from Tacoma to Carbonado, at the western base of the Cascade Mountains, a distance of thirty-four miles, and will eventually be extended across the range into eastern Washington. This division is also now under construction from the Columbia River westward through the valley of the Yakima, and has been surveyed and definitely located across the mountain range through Stampede Pass. From Tacoma the road passes through the Puyallup Valley, one of the most fertile spots known on the Pacific Coast, the seat of many hop yards, which realized for their owners large fortunes in the single crop of 1882, when hops sold at thirty-five cents to one dollar per pound. Amusing incidents are told of some of the hop growers, who had never had the comforts of life, and never been out of

Seattle, Puget Sound, W. T.

debt, and who suddenly found their cash balances swollen to sums they did not know how to take care of nor how to invest. These hop yards netted their owners in this single season from $10,000 to $60,000, and one enterprising speculator of the region is reported to have made $100,000. One worthy couple, who had been hitherto denied the ordinary conveniences of life, spent several days in examining the jewelry establishments of Tacoma to make heavy purchases. A bachelor of mature years invested in town property, bought more land, and then married. The more experienced of the Puyallup hop growers were enabled by this stroke of good fortune to begin valuable improvements upon their possessions. The pouring of not less than a quarter of a million dollars into a struggling neighborhood that had been fighting against fate for years, produced startling changes, indeed, in the entire community.

The stations passed are *Puyallup* (upon the grounds of the Puyallup Indian Reservation), nine miles; *Alderton*, thirteen miles; *Orting*, nineteen miles; *South Prairie*, twenty-six miles, and *Wilkeson*, thirty-two miles, respectively, from Tacoma. From near the latter station a spur track is laid to *Carbonado*, two miles farther.

The whole Sound region seems to be underlaid with coal beds. Wilkeson was the first point east of Tacoma that attracted attention, and mines were opened there years ago. Prospectors discovered the existence of a better quality of bituminous coal a few miles off, on the Carbon River, and capitalists of the Central Pacific Railroad Company became interested in developing the new mines. Heavy shipments of this coal are now made to San Francisco by way of Tacoma, at which point the Northern Pacific Railroad Company has built immense bunkers to facilitate traffic. The tourist who can spare time to visit Carbonado, a mountain mining town, and descend into the cañon of Carbon River, below it, will read on the abrupt walls graphic chapters of geology.

The Seattle Branch.—This line, built in the summer of 1883, to connect Tacoma with Seattle, leaves the main line at Puyallup, on the Cascade Division, and runs in a northerly direction over a low divide, which separates the streams that flow toward Tacoma on the south and Seattle on the north. It connects on White River with the railroad that is laid from Seattle to the productive coal mines at Renton and Newcastle, and runs down the valleys of the White River and the Dwamish to Elliot Bay. The length of this branch is twenty-nine miles, but there are no towns of consequence as yet along the route.

Seattle—(38 miles from Tacoma; population, 7,000.)—This is the largest city on Puget Sound, charmingly situated on and among high terraces, which rise steeply from the east shore of Elliot Bay. The city has lately grown with great rapidity, the assessed valuation of its property now exceeding $4,000,000. It is well built, having fine business blocks and handsome residences. The University of Washington Territory is established here, and there are schools, churches, three daily newspapers, bank buildings, hotels, saw and shingle mills and furniture factories. Three new wharves were built in 1882, and the wharf capacity has doubled since 1880. Over $100,000 were spent in 1882 in rebuilding the gas works and in laying new mains. The electric light also is largely in use. The suburbs of the city extend to the beautiful shores of Lake Washington, affording admirable sites for country seats and summer resorts. Seattle does a large business, principally in coal and lumber. Her extensive coal fields are the most productive on the coast, and considerable bodies of the richest and most available farming lands in the Territory lie near at hand. Very important coal mines are at Newcastle and Renton, near Seattle. The Newcastle mines are owned by the Oregon Improvement Company. Their product is a pure lignite, well adapted to household and railway purposes. The coal fields are connected by a railroad with Seattle, and are distant twenty-two miles from the city, whence the coal is shipped to San Francisco. These mines are now yielding nearly 800 tons a day, and are capable of producing 1,000 tons. The Oregon Improvement Company contemplates building a second freight wharf at Seattle, with warehouses capable of holding over 5,000 tons of grain; this, with the warehouses already existing, will give a total storage capacity of 8,000 tons. This Company has a fleet of four new steam colliers, each vessel averaging two and one-half trips a month between Seattle and San Francisco.

Seattle is also the principal port on Puget Sound for the fleet of large passenger steamships in the Pacific Coast trade. Its well sheltered harbor, entirely free from obstructions, affords good anchorage, and the water is deep enough for the largest vessels to lie alongside the wharves. There is also a great trade in salmon, and in manufactures of wood, flour and iron.

Distant View of Mount Tacoma.

The Glaciers of Mount Tacoma.

In the summer of 1882 the ice-fields of Mount Tacoma were visited by Mr. Bailey Willis, Assistant Geologist of the Northern Transcontinental Survey, who reported the existence of a group of glaciers which are said to equal those of Switzerland in size and grandeur.

The full extent of these glaciers has not yet been determined, but further explorations are in progress, and already an easy bridle-path has been made from Wilkeson, on the Cascade Branch of the Northern Pacific Railroad, to the foot of one of the more important glaciers, out of which springs the Carbon River. This bridle-path has been cut through the forest at the expense of the railroad company for the convenience of tourists. Guides, horses and all other requisites for a trip to the alpine region are to be found at Wilkeson. There is no doubt that the glaciers of Mount Tacoma and the magnificent scenery which environs them will attract large numbers of visitors, who now specially cross the ocean in order to see precisely those wonderful phases of creation which are to be found in equal sublimity at home. The accompanying pictures of the glaciers are from sketches made by Mr. Willis, from whose description of his journey to this new and interesting field of exploration the following extracts are also made:

"The Puyallup River, which empties into Puget Sound near Tacoma, heads in three glaciers on Mount Tacoma. During the summer months, when the ice and snow on the mountain are thawing, the water is discolored with mud from the glaciers, and carries a large amount of sediment out to Commencement Bay. If the Coast Survey charts are correct, soundings near the centre of the bay have changed from 100 fathoms and 'no bottom' in 1867, to eighty fathoms and 'gray mud' in 1877. But when the nights in the hills begin to be frosty, the stream becomes clearer, and in winter the greater volume of spring water gives it a deep green tint.

"For twenty miles from the sound the valley is nearly level. The bluffs along the river are of coarse gravel, the soil is alluvium, and a well sunk 100 feet at the little town of Puyallup passed through gravel and sand to tide mud and brackish water. From the foot hills to its mouth the river meanders over an old valley of unknown depth, now filled with material brought down by its several branches. About eighteen miles above its mouth the river forks, and the northern portion takes the name of Carbon River; the southern was formerly called the South Fork, but it should retain the name of Puyallup to its next division far up in the mountains. A short distance above their junction both Carbon River and the Puyallup escape from narrow, crooked cañons, whose vertical sides, 100 to 300 feet high, are often but fifty feet apart. From these walls steep, heavily timbered slopes rise 200 to 800 feet to the summits of the foot-hills. These cañons link the buried river basin of the lower stream with the upper river valleys. The latter extend from the heads of the cañons to the glaciers. They are apparently the deserted beds of mightier ice rivers, now shrunk to the very foot of Mount Tacoma.

"From Tacoma the entire course of the Puyallup and part of Carbon River are in view. Across Commencement Bay are the tide marshes of the delta; back from these salt meadows the light green of the cottonwoods, alder and vine-maple mark the river's course, till it is lost in the dark monotone of the fir forest. No break in the evergreen surface indicates the place of the river cañons; but far out among the foot-hills a line of mist hangs over the upper valley of Carbon River, which winds away eastward, behind the rising ground, to the northern side of Mount Tacoma. Milk Creek, one of its branches, drains the northwest spur, and on the western slope the snows accumulate in two glaciers, from which flow the North and South Forks of the Puyallup. These streams meet in a level valley at the base of three singular peaks, and plunge united into the dark gateway of the cañon.

"A trip to the grand snow peak from which these rivers spring was within a year a very difficult undertaking. There was no trail through the dense forest, no supply depot on the route. No horse nor donkey could accompany the explorer, who took his blankets and provisions on his back, and worked his way slowly among the towering tree trunks, through underbrush luxuriant as a tropic jungle. But last summer a good horse trail was built from Wilkeson to Carbon River, crossing it above the cañon, sixteen miles below the glacier, and during the autumn it was extended to the head of the Puyallup. Wilkeson is on a small tributary of Carbon River, called Fletts Creek, at a point where the brook runs from a narrow gorge into a valley about a quarter of a mile wide. The horse trail climbs at once from Wilkeson to the first terrace, 400 feet above the valley; then winds a quarter of a mile back through the forest to the second ascent of 100 feet, and then a mile over the level to the third. Hidden here beneath the thick covering of moss and undergrowth of the primeval forest, 1,400 feet above the present ocean level, are ancient shore lines of the sea, which has left its trace in similar terraces in all the valleys about the Sound. Thence the trail extends southward over a level plateau. Carbon River Cañon is but half a mile away on the west, and five miles from Wilkeson the valley above the cañon is reached. The descent to the river is over three miles along the hillside eastward.

"From Wilkeson to the river the way is all through a belt of forest, where the conditions of growth are very favorable. The fir trees are massive, straight and free from limbs to a great height. The larger ones, eight to twelve feet in diameter on a level with a man's head, carry their size upward, tapering very gradually, till near the top they shoot out a thick mat of foliage, and the trunk in a few feet diminishes to a point. One such was measured; it stands like a huge obelisk, 180 feet, without a limb, supporting a crown of but forty feet more. The more slender trees are,

curiously enough, the taller; straight, clear shafts rise 100 to 150 feet, topped with foliage whose highest needles would look down on Trinity spire. Cedars, hemlocks, spruce and white fir mingle with these giants, but they do not compete with them in height; they fill in the spaces in the vast colonnades. Below is the carpet of deep golden green moss and glossy ferns, and the tangle of vines and bushes that covers the fallen trunks of the fathers of the forest. The silence of these mountains is awesome, the solitude oppressive. The deer, the bear, the panther are seldom met; they see and hear first and silently slip away, leaving only their tracks to prove their numbers. There are very few birds. Blue jays, and their less showy gray, but equally impudent, consins, the "whiskey jacks," assemble about a camp; but in passing through the forest one may wander a whole day and see no living thing save a squirrel, whose shrill chatter is startling amid the silence. The wind plays in the tree tops far overhead, but seldom stirs the branches of the smaller growth.

"To the head of Carbon River from the bridge, on which the trail crosses it, is about sixteen miles. The rocky bed of the river is 100 to 200 yards wide, a gray strip of polished boulders between sombre mountain slopes, that rise sharply from it. The stream winds in ever shifting channels among the stones. About six miles above the bridge Milk Creek dashes down from its narrow gorge into the river. The high pinnacles of the spur from which it springs are hidden by the nearer fir-clad ridges. Between their outlines shines the northern peak of Mount Tacoma, framed in dark evergreen spires. Its snow fields are only three miles distant, but Carbon River has come a long way round. For six miles eastward the undulating lines of the mountains converge, then those on the north suddenly cross the view, where the river cañon turns sharply southward.

"Three miles from this turn is Crescent Mountain, its summit a semi-circular gray wall 1,000 feet high, the rim of a large crater. At sunset the light from the west streams across the head of Milk Creek and Carbon River, illuminating these cliffs as with the glow of volcanic fires, while twilight deepens in the valley. The next turn of the river brings Mount Tacoma again in view. Close on the right a huge buttress towers up, cliff upon cliff, 2,500 feet, a single one of the many imposing rock masses that form the Rugged Spur between Carbon River and Milk Creek. The more rapid fall of the river, the increasing size of the boulders, show the nearness of the glacier. Turning eastward to the south of Crescent Mountain, you pass the group of trees that hide it.

"This first sight is a disappointment. The glacier is a very dirty one. The face is about 300 feet long and thirty to forty feet high. It entirely fills the space between two low cliffs of polished gray rock. Throughout the mass the snows of successive winters are interstratified with the summer's accumulations of earth and rock. From a dark cavern, whose depths have none of the intense blue color so beautiful in crevasses in clear ice, Carbon River pours out, a muddy torrent. The top of the glacier is covered with earth about six inches deep, contributed to its mass by the cliffs on either side and by an island of rock, where a few pines grow, entirely surrounded by the ice river. The eye willingly passes over this dirty mass to the gleaming northeast spur of the mountain, where the sunlight lingers after the chill night wind has begun to blow from the ice fields.

"The disappointment of this view of the glacier leaves one unprepared for the beauty of that from Crescent Mountain. The ascent from a point a short distance down the river is steep, but not dangerous. The lower slopes are heavily timbered, but at an elevation of 4,000 feet juniper and dwarf pine are dotted over the grassy hillside. Elk, deer and white mountain goats find here a pleasant pasture; their trails look like well trodden sheep paths on a New England hill. The crest of the southwest rim of the crater is easily gained, and the grandeur of the view bursts upon you suddenly. Eastward are the cliffs and cañons of the Cascade Range. Northward forest covered hill and valley reach to Mount Baker and the snow peaks that break the horizon line. Westward are the blue waters of the Sound, the snow-clad Olympics and a faint, soft line beyond; it may be the ocean or a fog bank above it. Southward, 9,000 feet above you, so near you must throw your head back to see its summit, is grand Mount Tacoma; its graceful northern peak piercing the sky, it soars single and alone. Whether touched by the glow of early morning or gleaming in bright noonday, whether rosy with sunset light or glimmering ghost-like in the full moon, whether standing out clear and cloudless or veiled among the mists it weaves from the warm south winds, it is always majestic and inspiring, always attractive and lovely. It is the symbol of an awful power clad in beauty.

"This northern slope of the mountain is very steep, and the consolidated snow begins its downward movement from near the top. Little pinnacles of rock project through the mass, and form eddies in the current. A jagged ridge divides it, and part descends into the deep, unexplored cañon of White River, probably the deepest chasm in the flanks of Mount Tacoma. The other part comes straight on toward the southern side of Crescent Mountain, a precipice 2,000 feet high; diverted, it turns in graceful flowing curves, breaks into a thousand ice pyramids, and descends into the narrow pass, where its beauty is hidden under the ever-falling showers of rock.

"This rim of the crater you stand upon is very narrow; 100 feet wide, sometimes less, between the cliff that rises 2,000 feet above the glacier and the descent of 1,000 feet on the other side. Snow lies upon part of this slope; stones, started from the edge, leap in lengthening bounds over its firm surface, and plunge with a splash into the throat of the volcano, from which they had been thrown molten. A lakelet fills it now, and the ice slope, dripping into it, passes from purest white to deepest blue.

"A two days' visit to this trackless region sufficed only to see a small part of the magnificent scenery. White River Cañon, the cliffs of Ragged Spur, the northern slope of Mount Tacoma, where the climber is always tempted upward, might occupy him for weeks. Across the snow fields, where Milk Creek rises, is the glacier of the North Fork of the Puyallup, and the end of the horse trail we left at Carbon River is within six miles of its base."

The foregoing description of the Mount Tacoma glaciers may well be supplemented with the remarks made by Senator Geo. F. Edmunds, of Vermont, as printed in the Portland *Oregonian*, just after his return from a visit to the Carbon River glacier, in the company of Mr. T. F. Oakes,

Glaciers of Mount Tacoma.

Vice-President of the Northern Pacific Railroad, towards the latter part of June, 1883. The Senator said:

"I never believed there was anything in America comparable in grandeur to the scenery I have seen on my necessarily brief visit to Mount Tacoma. The access, thanks to the well cut trails made by the engineer of the Northern Pacific Railroad, is easy, and, as I learn, will be made as easy as a carriage-road within a month from now. To express half of my admiration for the transcendent grandeur of every part of the scenery, and especially of the glacier, would be impossible. A more perfect glacier, in all the features found in such a phenomenon, it would be impossible to find. Certainly no Alpine glacier excels it, and yet, as I was given to understand, it is the least in point of size of all that have been discovered on this mountain. I cannot help saying that I am thoroughly convinced that no resort in the United States will be so much sought after as this when once people come to know that what men cross the Atlantic to see can be seen in equal splendor, if not surpassed, at home.

"I hear doubts expressed as to whether Mount Tacoma can be ascended, but a steady head, a sure foot, a reliable alpenstock, and a little determination could probably accomplish what is, of course, a very daring feat. Tacoma has a fortune in the fact that it is the best point from which to start to this, the grandest of all American mountains; and I learn that guides and all conveniences will be afforded to tourists desirous of visiting this magnificent scene. If Switzerland is rightly called the playground of Europe, I am satisfied that around the base of Mount Tacoma will become a prominent place of resort, not for Americans only, but for the world besides.

"I need not deal with the particulars of our journey. Let me only add this much, that nothing could be, to me at least, more enjoyable than the gradual approach to the inner circle of the court where the monarch of our northern mountains reigns supreme. The emotions stirred in one's breast—at least they were in mine—completely defy all the powers of language to express. When we reached the foot of the glaciers, footsore and weary as we were, I could not help pausing in breathless silence as we reviewed this majestic mass of ice, imbedded in the bosom of this gigantic monarch among the alps. If it was not the grandest of the group, as we were informed by Mr. Willis, who was one of our party, it was enough magnificence for us. The stupendous sides of the cañon, in which the glacier lay, formed a setting perfect in its harmony of contrast, if the term be admissible, to the pagodas and pinnacles of the secluded mass of ice.

"I have been through the Swiss mountains, and I am compelled to own that, incredible as the assertion may appear, there is absolutely no comparison between the finest effects that are exhibited there and what is seen in approaching this grand isolated mountain. I would be willing to go 500 miles again to see that scene. This continent is yet in ignorance of the existence of what will be one of the grandest show-places, as well as a sanitarium."

The Lower Columbia River.

PORTLAND TO ASTORIA.—DISTANCE, 98 MILES.

A delightful excursion can be made from Portland to Astoria, and to the ocean entrance beyond. Fine steamers of the Oregon Railway and Navigation Company's fleet, affording excellent accommodations and good fare, run on this route. Leaving Portland in early morning, the steamer goes for twelve miles down the Willamette, giving opportunity to observe the progress which the city is making in its manufacturing and other enterprises. After an hour has passed the broader flow of the Columbia is reached. Now the busy commerce of the river is seen, as well as the farms and villages along the shores, and the tourist may look around on snowy peaks, whose summits overtop the mountains, and the sight of which never tires.

During summer and spring the vivid green of the "continuous woods where flows the Oregon," relieves the landscape of monotony. The shores and islands are tangled forests, thick with all native small growths, and the hillsides are covered with heavy fir. Sometimes rock ledges crop out, but the Lower Columbia has few of the startling effects peculiar to the gorge it has cleft through the Cascades. Passing Sauvie's Island, the towns of St. Helen's, Columbia and Ranier, we get into the region of the Coast Mountains, and see where the river has worked its way through these also, but not so abruptly as through the much wider and loftier Cascade range. The lower river is, nevertheless, beautiful, with its wilderness of shores and islands; its occasional bluffs and cliffs; its broad flood, sometimes widening to miles, and always spreading out in majestic volume. Enjoying all the comforts of first-class travel, the tourist reads his book, or pencils his notes, as the boat passes down the stream, whose long reaches reveal many attractions for the eye to dwell upon.

Mount Coffin is an island rock, near the Oregon shore, not far below Columbia City, on which grew a straggling forest years ago, and which was famous when the first settlers came to Oregon as an Indian place of sepulture. The red men came hither from a distance, usually placing the body in a canoe that had belonged to the dead, and filling it with his utensils and equipments. The canoe was rendered useless, the bow and arrows broken, and the effects were spoiled, by breaking or other injury, before the deceased was laid away to rest in the canoe, suspended among the limbs of the trees. It is easily to be imagined that Mount Coffin was reserved as the resting place of noted Indians, and that long fleets of canoes and much superstitious ceremonial attended the obsequies. Twenty-five years ago there were many vestiges of this Indian necropolis, but now the aborigines are nearly extinct, and vandal hands have despoiled this primitive cemetery.

Towards Astoria *Tongue Point* protrudes into the river from the Oregon side. It is a notable landmark, whereon the Government should place a beacon or light-house. Its recent owners say that there are twenty acres of level land on its summit. Below Tongue Point the river broadens to five miles, and Astoria faces the downward passage from the eastern slope of *Clark's Point*—named after the early explorer—which juts seaward, with the river on one side and Young's Bay towards the ocean. Beyond it, again, is Clatsop Peninsula, reaching twenty miles south of the entrance to the Columbia River, with Young's Bay on the east and the Pacific Ocean on the west.

The business part of *Astoria* is built upon piles, and its residences climb the sides of the ridge, which protects it from southwest storms. The prosperity of the place depends considerably on the local commerce of the country, but it is practically the seaport of Oregon. Here great ships anchor before they go up the river, and here they often wait to complete their cargoes. The broad bay is alive with shipping. The wharves of Astoria show all sorts of river and ocean craft, especially tugs and pilot boats peculiar to a seaport of consequence. The town grows rapidly and must grow more rapidly, for the commerce of the whole region pays it some measure of tribute. It has now, perhaps, an immature appearance, which, however, is fast passing away, as substantial business structures and comfortable private residences are rising in every direction.

Lumbering is becoming an important industry on the lower river, but the great incentive to growth at Astoria is the salmon trade and fisheries. These already include over fifty great canning establishments, that employ thousands of men during the fishing and packing season. The shores above and below the city are lined with these salmon packing establishments, and the traveler of an inquiring turn of mind can find much to interest (rather than to enjoy) in watching the processes by which the great fish are prepared for market. There are various kinds of salmon in these western rivers, and it is a singular fact that fish with well-marked peculiarities confine themselves to the different streams. Some of the salmon family which are found in the rivers of Oregon and Washington are not especially valuable, though they would no doubt be appreciated highly where the fish is not so plentiful. The king of edible fishes is the "chinook salmon," of the Columbia. It runs from April 1st to August 1st of each year, although other varieties run during other months. It is for this variety of salmon that thousands of boats are equipped with deep-water nets, and venture to the very edge of the surf to secure the fish, if possible, before it reaches fresh water. It is a reckless business, and occasionally causes loss of life, for sudden changes of wind may drive boats and nets into the breakers. A most beautiful sight on the lower river is to see the fleet of fishing boats, in squadrons, make their way before the wind, or beat down with many a tack, with full sails spread, to the fishing ground. The best fishing is in the night time, and the boats go down in the evening and return in early morning. From the balcony of the hotel at Astoria, or from some good point of view on the hill, a fine chance to see this fleet of boats sailing down the bay, or coming home with the spoils of their night's work, is offered. Astoria is over twenty miles from the bar, and the broad bays between make it an especially beautiful harbor. The fringe of surf on the outer bar indicates plainly the channels from the ocean to the inner bay.

From Astoria delightful excursions may be made in different directions. Young's River, at the head of Young's Bay, has a remarkable water-fall that has been often sketched. Parties go in small steam tugs to visit this forest-hid cascade. Across *Clatsop Plains*, some twenty miles from Astoria, by land and water, is a favorite resort known as the "Seaside," where good hotels are found for summer boarders. On one side a fine ocean beach, and on the other a trout creek comes down from the Coast Mountains, where elk and deer abound. Seventy or more years ago the explorers, Lewis and Clark, came down the Columbia and wintered on these plains, because there were droves of almost tame elk there, and meat was abundant. The small stream on which they camped is named Lewis and Clark River to this day.

Over on the north shore of Baker's Bay, near the ocean, is *Ilwaco*, another watering place. Many Portlanders have built small camps there, or on the ocean beach near by, and spend the warm weeks of summer enjoying the cool breath of the Pacific Ocean and bathing in the pounding surf.

The bold headland that forms the cape north of the entrance, on the Washington side, is now called Cape Hancock, although it was long known as Cape Disappointment. Under the hill, inside the bay, are the military buildings and residences of the officers that command the fort on the heights. The view from the cape is very fine. The earthworks have great columbiads pointing seaward, and the presence of a light-house lends interest to the spot. The outlook shows a sweep of the outer ocean and the inner bay, and you can see the surf dashing on the winding shore line to the north. This picture is reproduced on the great lenses of the light-house with such vivid effect as to itself repay a journey to the spot. Fort Stevens, on Point Adams, at the north of Clatsop Peninsula, is a low earthwork to defend the south entrance. These are all places of interest, and worth visiting.

Sound and Ocean.

TRIPS ON PUGET SOUND, TO SAN FRANCISCO AND TO ALASKA.

Excursions on Puget Sound.—The steam-boats of the Oregon Railway and Navigation Company ply between all the principal ports on Puget Sound. These vessels are well adapted to the passenger service in which they are engaged, and a trip upon the waters of the "Mediterranean of the Pacific" is to be highly recommended. Puget Sound covers an area of over 2,000 square miles, and has a shore line of 1,800 miles. The dark blue water, free from reefs or shoals, is so deep that the largest sea-going vessels may approach the bold shores at almost every point. Islands are so many that the Sound is in reality a beautiful archipelago, extending from the British line on the north, and embracing the Straits of San Juan de Fuca, which afford a broad and unobstructed channel to the ocean, the Gulf of Georgia, the Canal de Haro, Bellingham Bay, Rosario Straits, Possession Sound, Admiralty Inlet, Hood's Canal and other passes and estuaries. There is here a labyrinth of lakes, rivers, peninsulas and islands, the vision being bounded westward by the snow crowned Olympic Range, while the grand white peak of Mount Baker, in the Cascade Range, looms up to its height of 11,100 feet toward the north, the whole forming a succession of pictures of great diversity and beauty. The shores and heights are heavily wooded, some of the giant firs being the longest and straightest anywhere to be found, and much in request for vessels' masts.

The steamers run daily from Olympia *via* Tacoma to Seattle. From Tacoma they make tri-weekly trips to the great lumber mill villages at Ports Gamble and Ludlow, thence to Port Townsend, a military post and the port of entry to the United States, and then across the Straits of Juan de Fuca, to Victoria, on Vancouver Island, distant 155 miles from Tacoma. Victoria is the pleasantest and most attractive city in British Columbia, with broad ocean and grand mountain views. Three miles distant from Victoria is the fine harbor of Esquimalt, where there is a dockyard and supply station for Queen Victoria's vessels on the Northern Pacific. From Seattle there is bi-weekly communication with the islands of the Archipelago de Haro, and as far northward as Sehome, on Bellingham Bay, a distance of 121 miles. These are all delightful excursions.

Portland to San Francisco.—Between Portland and San Francisco there is direct communication at least twice a week by means of the steam-ships of the Oregon Railway and Navigation Company. The distance from port to port is 670 miles, and the time usually occupied by the voyage is from fifty-five to sixty hours. These steam-ships, in point of size, speed, accommodation and equipment, are equal to the finest vessels afloat.

Excursions to Alaska.—Every summer, in the early days of June, July and August, the Pacific Coast Steam-ship Company dispatches one of its fine vessels from San Francisco to Alaska, touching at Astoria and Portland, Oregon; Port Townsend, Washington Territory; Victoria and Nanaimo, British Columbia, and at Wrangle, Sitka and Juneau, Alaska. The round trip from San Francisco and return occupies from twenty to twenty-five days. This excursion is extremely interesting. It affords an opportunity to see the bold and mountainous line of the coast, the fine scenery of the Lower Columbia and Willamette rivers, the picturesque region of Puget Sound, and the novel sight of icebergs and glaciers on the voyage through the inland seas of the Alaskan archipelago. It is no longer necessary to visit Norway in order to see the sun at midnight. This may now be done at far less cost of time and trouble by making the trip to Alaska upon steam-ships which are especially fitted out with a view to the greatest comfort and entertainment of tourists. The price of round trip tickets, including berths and meals on the steam-ship, is $125 from San Francisco to Alaska and return, and $95 from Portland and back.

THE YELLOWSTONE NATIONAL PARK. A manual for tourists, with twenty-four illustrations, together with route, maps, and a plan of the Upper Geyser Basin. Also an appendix containing railroad lines and rates, as well as other miscellaneous information. By HENRY J. WINSER.

"An excellent guide."

"It should be of great value."

"A highly interesting little volume."

"Replete with interesting and valuable information."

"It is admirably gotten up, and conveys a very clear idea of that attractive wonderland. Altogether it is a valuable and instructive compilation."
—*Puget Sound Argus.*

"The book contains vivid descriptions of the wonderful hot springs, beautiful lakes, grand canyons, and sombre mountains of the Yellowstone Park."—*Seattle Chronicle.*

"The most complete work of its kind that we have seen. Contains every item of information that the tourist could ask for, and is, in fact, an indispensable hand-book. Mr. Winser has been in the regions described, and speaks of the things he has seen."—*Chicago Tribune.*

"The author has presented in a well-arranged form concise and graphic descriptions of the principal objects of interest, together with the necessary information as to the best methods of reaching them. Not only to tourists contemplating a trip to the Park, but to all admirers of the wonders of nature, the little work will prove far more attractive than the average of its class."—*N. Y. Evening Post.*

"I deem your admirably arranged, finely illustrated and printed guide-book not only the first, worthy of the name, which has been published, but also, from your personal knowledge, painstaking quotations, and candid descriptions, a work which, if kept up with the march of future improvements, cannot be superseded."—P. W. NORRIS.

Price, in paper, **40 cents.**

G. P. PUTNAM'S SONS, Publishers,
27 & 29 West 23d Street, New York.

www.ingramcontent.com/pod-product-compliance
Lightning Source LLC
Chambersburg PA
CBHW020242240426
43672CB00006B/607